Oil Horror

Oliver Haiste

First published in Great Britain in 2008
Bloomsbury International

Copyright © 2008 by Oliver Haiste

ISBN: 978-0-9559847-0-9

Bloomsbury International
42 Soho Square
London
W1D 3QY

To Carrie

All My Love

CONTENTS

CONTENTS

Introduction

"Perhaps the sentiments contained in the following pages are not yet sufficiently fashionable to procure them general favor; a long habit of not thinking a thing wrong gives it a superficial appearance of being right, and raises at first a formidable outcry in defense of custom. But the tumult soon subsides. Time makes more converts than reason."
-Thomas Paine, The Common Sense (1776)

"It's hard to envisage the effects of a radically reduced oil supply on a modern economy or society. The implications are mind-blowing."
Michael Meacher. former UK Environment Minister

If one accepts that oil is a finite resource, then it follows that one must also accept that the production of oil will peak at some point. All that is therefore questionable is the date when oil will peak and whether or not alternatives are possible.
Oil geologists predict a peak before 2015, but mass market alternatives are not being developed anywhere near fast enough. At some point, global oil production will reach a peak and when it does, it could start a global economic and social crisis on a scale never seen before.

Peak Oil refers to the point at which new oil production can no longer keep up with declining oil fields, and this results in a yearly decline in the amount of oil produced. Since the world has become increasingly dependent upon petroleum year after year, declining petroleum production has the potential to severely disrupt our lives through much higher prices and fuel shortage.

We have allowed oil to become vital to virtually everything we do. Ninety percent of all our transportation whether by land, air or sea, is fuelled by oil. Ninety-five percent of all goods in shops involve the use of oil. Ninety-five percent of all our food products require oil use.[i] Just to farm a single cow and deliver it to market requires six barrels of oil, enough to drive a car from New York to Los Angeles.[ii] The world consumes more than 80 million barrels of oil a day, 29 billion barrels a year, at the time of writing. This figure is rising fast as it has done for decades. The almost universal expectation is that it will keep doing so for years to come. The US government assumes that global demand will grow to around 120 million barrels a day 43 billion barrels a year by 2025.[iii] The International Energy Agency, the organization set up by industrialized countries to give them advice on oil and other energy matters, is scarcely less bullish. Its 2004 Energy Outlook forecasts 121 million barrels a day by 2030.[iv] Few question the feasibility of this requirement, or the oil industry's ability to meet it. They should, because the oil industry won't come close to producing 120 million barrels a day. The most basic of the foundations of our assumptions of future economic well-being is rotten. Our society is in a state of collective denial that has no precedent in history, in terms of its scale and implications.

Of the current global demand, America consumes a quarter. Because domestic oil production has been falling steadily for thirty-five years, with demand rising equally steadily, America's relative share is set to grow, and with it her imports of oil. Of America's current daily consumption of 20 million barrels, 5 are imported from the Middle East, where almost two-thirds of the world's oil reserves lie in a region of especially intense and long-lived conflicts. Every day, 15 million barrels pass in tankers through the narrow Straits of Hormuz, in the troubled waters between Saudi Arabia and Iran.[v] The US government could wipe out the need for all their 5 million barrels, and staunch the flow of much blood in the process, by requiring its domestic automobile industry to increase the fuel efficiency of automobiles and light trucks by a mere 2.7 miles per gallon. But instead it allows General Motors and the rest to build ever more oil-profligate vehicles. Many sports utility vehicles (SUVs) average just 4 miles per gallon. The SUV market share in the US was 2 percent in 1975. By 2003 it was 24 percent. In consequence, average US vehicle fuel efficiency fell between 1987 and 2001, from 26.2 to 24.4 miles per gallon. This at a time, when other countries were producing cars capable of up to 60 miles per gallon.[vi]

Most US presidents since the Second World War have ordered military action of some sort in the Middle East. American leaders may prefer to dress their military entanglements east of Suez in the rhetoric of democracy building. Although democracy building is to varying degrees a motivation, the long-running strategic theme is obvious. It was stated most clearly and paradoxically, by the most liberal of them. In 1980 -Jimmy Carter declared access to the Persian Gulf "a vital national interest to be protected by any means necessary including military force". [vii]

This the USA has been doing ever since, clocking up a bill measured in the hundreds of billions of dollars and counting.

With such a strategy comes an increasingly disquieting descent into moral ambiguity, at least in the minds of something approaching half the country. The nation that gave the world such significant landmarks in the annals of democracy as the Marshall Plan is forced by its deepening oil dependency into a foreign-policy maze that involves arming some despotic regimes, bombing others, and scrabbling for reasons to make the whole construct hang together.

America is not alone in her addiction and her dilemmas. The motorways of Europe now extend from Clydeside to Calabria and Lisbon to Lithuania. Agricultural produce that could have been grown for local consumption rides needlessly along these arteries the length and breadth of the European Union. The Chinese attempt to emulate this mode even as they enforce production downtime in factories because of diesel shortages and despair that their vast national acreage seems to play host to so little oil.

This half-century of deepening oil dependency would be difficult to understand even if oil were known to be in endless supply. But what makes the depth of the current global addiction especially bewildering is that, for the entire time we have been sliding into the trap we have known that oil is in fact in limited supply At current rates of use the global tank is going to run too low to fuel the growing demand sooner rather than later this century. This is not a controversial statement, It is just a question of when. One purpose of this book is to explain why.

Why, then, have we not been seeking an earlier transition to the alternatives that must lie beyond oil dependency? Hydrogen fuel, biofuel, fuel cells and advanced batteries are among the technologies that can provide the direct power for transportation in the future.

Solar and many other forms of alternative energy can provide the electricity to split water into hydrogen and charge batteries. This too we have understood for decades. We have also known that there are massive untapped reservoirs of oil savings in energy-efficiency measures and innovative mass transit. These alternatives may not be able to replace oil quickly or easily, given their tiny current markets. But they work, and in most cases they have been waiting for a green light for years. This is without any further fruits of human ingenuity, suitably directed. In a society that put a man on the moon more than three decades ago, surely there can be no doubt that we could replace oil use if we seriously wanted to? I ask again, why have we not been fast-tracking the solutions to the problem long since? A second purpose of this book is to examine that question.

A third purpose of the book is to pose and endeavour to answer the question of how fast oil is now depleting. Finite resource that it is, there will come a day, inevitably, when we reach the highest amount of oil that can ever be pumped. Beyond that day; which we can think of as the topping point, or "peak oil" as it is often called, will lie a progressive overall decline in production. Putting the same question a different way then, at the current prodigious global demand levels, where does its topping point lie?

THE LATE TOPPERS VERSUS THE EARLY TOPPERS

A great battle is raging today, largely behind the scenes, about when we reach the topping point, and what will happen when we do. In one camp, those I shall call the "late toppers", are the people who tell us that 2 trillion barrels of oil remain to be exploited in oil reserves and reasonably expectable future discoveries. This camp includes almost all oil companies, governments and their agencies most financial analysts and most business journalists. As you might expect, given this line-up the late toppers hold the ascendancy in the argument as things stand. In the other camp are a group of dissident experts who I shall call the "early toppers". They are mostly people who have worked in the heart of the oil industry. The majority of them geologists many of them members of an umbrella organization called the Association for the Study of Peak oil (ASPO).They are joined by a small but growing number of analysts and journalists.

The early toppers reckon that 1 trillion barrels of oil or less are left.

In a society that has allowed its economies to become geared almost inextricably to growing supplies of cheap oil, the difference between 1 and 2 trillion barrels is seismic. It is roughly the difference between a full Lake Geneva and a half-full one, were that lake full of oil and not water. If 2 trillion barrels of oil or more indeed remain the topping point lies far away in the 2010s. The "growing" and "cheap" parts of the oil-supply equation are feasible until then at least in principle' and we have enough time to bring in the alternatives to oil. If only 1 trillion barrels remain however, the topping point will arrive some time soon and certainly before this decade is out. The growing and cheap parts of the oil-supply equation become impossible, and there probably isn't even enough time to make a sustainable transition to alternatives.

The worlds of economics and business routinely assume a future in which oil is in growing and cheap supply.

Economists tend to assume that their "price mechanism" will apply. Higher prices will lead to more attractive conditions for exploration. This will lead to more oil being found, and the inevitable discoveries will bring the price down until the next cycle. Massive corporations write five-year plans based on assumed access to cheap oil and gas. Think, for example, how important such access must be to a chemical company dealing in plastics derived from oil.

 Or a food-processing company reliant on oil for every stage of food transportation, including of perishable final products, plus almost all the bottling and packaging and many of the preservatives and additives. But suppose the economists and corporate planners are wrong? Imagine the collapse of confidence when a critical mass of financial analysts, across the full breadth of sectors in a stock exchange, conclude that they are wrong?

If the topping point is indeed imminent, economic depression looms as a real prospect. The Saudis were right to be scared of this possibility in the 1970s. In the (Great Depression of the 1930s, triggered in 1929 by the worst-ever stock-market crash, economic hardship was horrific world trade fell by a breathtaking 62 percent between 1929 and 1932.

The widespread unemployment and social unrest bred Fascism in many countries, in some nations on a scale that would change the course of human history.

As for the stock markets, it took them fifty years to regain their pre-collapse value in real terms.

8

As we shall see the consequences of oil depletion could prove catastrophic for the world population in myriad ways. It is the

lifeblood of virtually every economic transaction and the world is no-where near prepared to cope without it.

Chapter 2

The oil industry and the future of the world economy in a nutshell.

"That's it. I can now refer to the world oil peak in the past tense. My career as a prophet is over. I'm now an historian".

Kenneth S Deffeyes, Professor Emeritus of Geology, Princeton University, February 2006 [1]

This chapter is presented as a breakdown of the facts we know about the oil industry. I will provide information as succinctly and as objectively as possible, with no editorial analysis or personal commentary. I will do this in subsequent chapters.
It is important to reiterate that peak oil is not the point at which we totally run out of oil. There is often confusion on this issue. Although there is general consensus that we could be in trouble when oil totally runs out, many are still unaware that the trouble will come a long time before then, at the peak of production,
Peak oil is a term that is used to describe the point in time when the maximum rate of global petroleum production is reached, after which the rate of production enters into terminal decline.
There is very little disagreement about the notion that oil production will peak; the main focus of dispute is now around the timing of that peak. [2]
In the event that demand is higher than anticipated, the peak will come sooner than predicted, to be followed by a steeper decline.[3]
Similarly, reducing the demand for oil will delay the occurrence of peak oil. The Chief Executive of Total estimates that if oil demand growth were halved, the global oil peak would be delayed by a decade.[4]

Hubbert

Marion King Hubbert was a geophysicist (1903-1989) who suggested that the fossil fuel era would be "but a pip" in human history, with the consumption of energy from fossil fuels:

"rising sharply from zero to a maximum, and almost as sharply declining, and thus representing but a moment in the total of human history"[5].

He developed a theory, now called Hubbert peak theory, to accurately predict in 1956 that oil production in the United States of America would peak between 1965 and 1970[6] (the peak occurred in 1970.[7] His model has since been used to predict the peak petroleum production of many other countries. According to the model, the production rate of a limited resource follows a roughly symmetrical bell-shaped curve based on the limits of exploitability and market pressures.
The shape of the curve has since been shown to vary considerably[8], sometimes with several peaks, depending on economic, technical and political factors[9]. The concept of peak oil states that global production will follow a similar bell-shaped curve, with total production falling as oil field production peaks.

The Theory

The theory behind the exploitation of any finite, non-renewable resource is relatively simple.

Hubbert's theory states that:

Production starts at zero
Production rises to a peak;
Once the peak has been passed, production declines until the resource is depleted. [10]

The point of maximum production tends to coincide with the midpoint of depletion of the resource, which means that when the Hubbert peak is reached, approximately half of the recoverable oil available on earth will have been used. [11]

The best approximation of total maximum global oil production (estimated ultimate recovery) can be calculated by summation of: [12]

1) The amount produced to date (cumulative production)

2) Estimates of how much remains to be produced from known fields (reserves)

Several attempts have been made to calculate the estimated ultimate recovery. Most attempts also extrapolate a date for peak oil.

12

Estimated ultimate recovery and date of peak oil [13]

Bartlett[14]   2,000 billion 2004
Campbell [15] 1,925 billion 2010
Deffeyes [16] 2,013 billion 16 December 2005
Robelius[17] - Year 2018
Skrebowski [18] Year 2011
USGS[19]/EIA[20] 3,021 billion 2037

The Energy Information Administration in the USA estimates that world production will peak at 146 million barrels per day in 2037 [21].

The National Petroleum Council notes that oil companies' average projection of global petroleum production is 96 million barrels per day in 2015 and 107 million barrels per day in 2030 [22]. During 2005 (the year of historical maximum production), an average of 74 million barrels per day was produced .[23]

The alternative to attempting to calculate when peak oil will occur is to assume that it has already happened [24]. In this case, information from the Energy Information Administration indicates that peak oil passed in February 2008 [25], with 74.6 million barrels per day. 2005 was the year with the highest production (an average of 73.8 million barrels was produced per day [26]).

The principal problem with this approach is that the peak will only be known after the event, so it is not a useful predictive mechanism. Also, although production in 2007 was 0.7 per cent below the 2005 record, results from the first three months of 2008 indicate that production for the year may exceed the 2005 level.

The National Petroleum Council disagrees with the contentions of peak oil theorists.

Its main argument is that new discoveries of oil, and new technology for enhancing recovery from existing reservoirs and unconventional sources, will delay the onset of peak oil until the 2030s. [27]

3. The Political Background

David Goodstein, Professor of Physics and Applied Physics, California Institute of Technology. [29]

"Our present national and international leadership is reluctant even to acknowledge that there is a problem... the crisis will occur, and it will be painful. The best we can realistically hope for is that when it happens, it will serve as a wakeup call, and will not so badly undermine our strength that we are unable to take the giant steps that are needed".

3.1 The US Geological Survey

The US Geological Survey (USGS) produced a report in 2000 that suggested the global estimated ultimate recovery to be 3,021 billion barrels. [30]

However, this was based on assessments of new discoveries that have been challenged by many commentators. [31]

In particular, the report forecasted that a mean of 674 billion barrels would be found between 1995 and 2025, an average of 23 billion barrels per year. [32]

Between 1995 and 2002, the average of new discoveries was 10 billion barrels per year. [33] Historically, the largest oil fields tend to be discovered first; in mature oil provinces, progressively smaller fields are discovered [34]. The fact that actual discoveries have departed so far from the USGS modelling suggests an overestimate of the likely ultimate recovery.

3.2 The International Energy Agency

The International Energy Agency (IEA) acts as energy policy adviser to 27 member countries, including all 15 pre-enlargement EU Member States. It aims to assist its members' efforts to"ensure reliable, affordable and clean energy for their citizens".[35]

The IEA was founded during the oil crisis of 1973-74 as a result of an agreement on an International Energy Programme. Its initial role was to co-ordinate mitigation measures in times of oil supply emergencies. The objectives of the Agreement are: [36]

To maintain and improve systems for coping with oil supply disruptions.
To promote rational energy policies in a global context through co-operative relations with non-member countries, industry and international organisations.
To operate a permanent information system on the international oil market.
To improve the world's energy supply and demand structure by developing alternative
energy sources and increasing the efficiency of energy use.
To promote international collaboration on energy technology.
To assist in the integration of environmental and energy policies.

The Agreement requires IEA member countries to hold oil stocks equivalent of at least 90 days of net oil imports and, in the event of a major oil supply disruption, to release stocks, restrain demand, switch to other fuels, increase domestic production, or share available oil, if necessary.
Such measures would not be taken in response to price increases, but rather an actual physical shortage. [37]

15

Public stocks of oil, held exclusively by member countries for emergency purposes, were 1.5 billion barrels in mid 2007.[38] At a drawdown rate of 4 million barrels per day, these stocks would last one year. The most recent supply disruption of this scale occurred during the Iraq invasion of Kuwait (production of 4.3 million barrels per day was lost from August 1990 to January 1991).[39]

This was one of only two events during which the IEA has acted to bring additional oil onto the markets (the other was in response to the hurricanes in the Gulf of Mexico in 2005, which caused production losses of 1.5 million barrels per day).[40]

The IEA considers world resources to be sufficiently abundant "to sustain likely growth in the global energy system for the foreseeable future".[41] However, it tempers this assessment with uncertainty as to whether sufficient investment will be committed to new output up to 2030 in order to compensate for falling output in existing fields, and to keep pace with the projected increase in demand.[42]

3.3 The oil industry

Much of the oil industry contests the notion that oil production will peak in the near future. For example, the National Petroleum Council in the USA suggests that:

"the world is not running out of energy resources, but there are accumulating risks to continuing expansion of oil and natural gas production from the conventional sources relied upon historically".[43]

16

ExxonMobil placed a one-page advert in the New York Times on 2 March 2006 (in response to a peak oil article [44]) stating that:

"Oil is a finite resource, but because it is so incredibly large, a peak will not occur this year, next year, or for decades to come... conservative estimates of heavy oil and shale oil push the total resource well over four trillion barrels" [45].

ExxonMobil's oil production fell almost 10 per cent in the first three months of 2008. [46]

The oil industry forecasts for global oil production are higher than the lowest estimates (usually those provided by peak oil theorists), but lower than the EIA and IEA reference case scenarios. [47]

Oil companies are required to list their reserves with the US financial market regulator Securities and Exchange Commission (SEC) in the USA. The SEC regulates share trading to ensure that investors are not defrauded. An oil company's value is based largely on its reserves, and this is reflected in its share price. [48] On 9 January 2004, Royal Dutch/Shell announced that 20 per cent of its 'proved' reserves [49] should be recategorised (downgraded). [50]

Following the admission, £3 billion was lost on the share value [51] and the company paid more than £200 million in fines and compensation. [52]

Shell recognises the inevitability of peak oil. In an internal email to staff, the Chief Executive noted that:

"Shell estimates that after 2015 supplies of easy-to-access oil and gas will no longer keep up with demand". [53]

17

The President of Shell has also commented that the recent
high price of oil is related to a lack of supply:

"The fundamental laws of supply and demand are at work."[54]

In May 2008, the Chairman and President of BP America Inc
stated:

*"We cannot change the world market... Today's high prices are
linked to the failure both here and abroad to increase supplies,
renewables and conservation."* [55]

3.4 Industry Commentators

Yergin notes that:

*"This is not the first time that the world has 'run out of oil'. It's
more like the fifth. Cycles of shortage and surplus characterize
the entire history of the oil industry"*.[56]

The article was, however, written when oil was at $60 per barrel,
in 2005. The commentator's record on predicting oil prices has
itself been criticised by others [57].

Riva cautions that:

*"The power of the quest for profits is sometimes
underappreciated by geologists in their projections of future oil
availability... An oil supply crisis in the next few years appears
somewhat premature, but the scales seem to be tipping in favor
of geology. In the context of the 21st century, a world oil supply
crisis, while not imminent, appears more likely sooner than
later"*.[58]

In 1999, at historically low oil prices, Udall and Andrews noted:

"Although it's hard to imagine sharply higher oil prices during an oil glut, what seems unlikely is inevitable. The crunch may arrive suddenly or in slow motion".

As former Energy Secretary Don Hodel says:

"We're sleepwalking to disaster." When it happens, journalists will shout, "We're running out of oil." That's not true. Rather, we are running out of cheap oil. After production peaks oil will be readily available at a higher price, though in declining amounts, for at least 50 years. What we face is not a short-term crisis but a chronic shortfall... the transition to more expensive oil could be bumpy".[59]

Other commentators are convinced that peak oil will manifest itself shortly, or that it has already occurred. Deffeyes notes:

"The profits of major oil companies are piling up by the tens of billions of dollars per quarter. They are hoarding cash, buying back stock, and declaring dividends. They are not investing heavily in new facilities. If oil production has ceased growing and is about to decline, nobody needs new refineries, new pipelines, or new tanker ships. Most telling of all, the majors are not increasing their investment in exploration drilling."[60]

The lack of investment is affirmed by Alan Greenspan, former US Federal Reserve Chairman, who claimed that companies have invested too little in production and infrastructure to keep supply growing in line with higher demand.[61]

Gordon Brown, the UK prime minister, expressed the hope that some of BP and Shell's record £7 billion first-quarter earnings in

2008 would be invested in "getting more oil out of the North Sea"[62].

In May 2008, the US Energy Secretary commented that:

"The high-priced energy environment is being driven by the fact that demand has outstripped supply"[63],

The European Commission noted in June 2008 that:

"The current surge in oil prices is largely the result of a major structural shift in oil supply and demand in the global economy. Oil supply is struggling to keep pace with rising global demand".[64]

Simmons refers to the importance of 'giant' oilfields (those producing more than 100,000 barrels a day) in global oil supply. Approximately 3 per cent of the world's 4,000 active oilfields produce 47 per cent of the world's supply.[65] The 14 largest oil fields produce 20 per cent of the world's supply.

The average age of these fields is 44 years.[66] None of the new fields discovered recently is projected to provide daily production in excess of 250,000 barrels (0.34 per cent of global production).[67]

3.5 The Energy Information Administration

The Energy Information Administration (EIA) is the US Government provider of energy statistics.
In May 2007 it published its International Energy Outlook 2007 (IEO 2007).[68]

The IEO 2007 is based around a number of scenarios.

Under the reference scenario of IEO 2007, world consumption of petroleum[73] increases from 83 million barrels per day in 2004 to 97 million in 2015 and 118 million in 2030.[74]

Such a projection does not recognise the possibility either that peak oil has already occurred (at a maximum of 74
million barrels per day in 2005), or that it will occur in the near future.

The series of International Energy Outlook documents produced since 2001 has consistently underestimated the likely future price of oil by between 60 and 91 percent.

The EIA information uses industry standard journals and sources, and is regarded as being one of the main repositories of information on the subject. According to the EIA, proved reserves[75] of global oil are approximately 1.2 trillion barrels.[76]

However, these proved reserves include the official reports from the member countries of the Organisation of the Petroleum Exporting Countries (OPEC).

Many commentators outside the oil establishment regard the

reserve reporting from some of the OPEC countries as dubious.[77] One example of mistrust in official figures derives from the Iraqi troops' firing of Kuwaiti oilfields, which destroyed 2 billion barrels of oil, but which was not reflected in official Kuwaiti reserve figures.[78]

In January 2006, Petroleum Intelligence Weekly uncovered official data that suggested Kuwait's reserves to be less than half the official value.[79]

More significantly, during the 1980s, there was discussion that the size of OPEC countries' reserves should be a factor in the calculation of a country's export quota. Since a larger export quota meant greater oil export revenue for the country concerned, there is suspicion that the increases in OPEC quotas between 1986 and 1991 (amounting to more than 300 billion barrels of oil) were little more than a political exercise[80]. Growth in reserves over the period is believed to have contributed no more than 100 billion barrels to OPEC inventories[81]. Furthermore, between 1990 and 1999, there were no significant changes in the annually claimed oil reserves of seven of the largest oil producers in the world[82], even though they produced 105 billion barrels of oil between them (this amount was not deducted from their 'proven reserves').[83] There is no way to audit sovereign governments' claims.[84]

3.6 The UK Government

The UK Government has argued that oil (and gas) resources will not peak for at least a few decades. In 2006, it was the UK Government's view that although global oil and gas production will one day peak:

"We believe that such a peak is not imminent and will not be reached until some time after 2030, provided the necessary

22

investments in expanding and replacing production capacity are made"[85].

In October 2007, the UK Government's assessment was that:

"the world's oil and gas resources are sufficient to sustain economic growth for the foreseeable future".[86]

Then, in May 2008, the UK's prime minister stated that:

"The cause of rising prices is clear: growing demand and too little supply to meet it both now and perhaps of even greater significance - in the future... Our strategic interests... all now point in the same direction: decreasing dependency on oil"[87].

The UK Government's analysis of the oil supplies within its own jurisdiction has proven erroneous. In the 2003 Energy White Paper, the UK Government predicted that "by around 2006 we will also be a net importer of gas and by around 2010 of oil".[88] The UK became a net importer of gas in 2004[89], and of oil in 2006.[90]

4 . Oil Prices and Production

Dr M King Hubbert (1903-1989), Former Senior Research Geophysicist at the US Geological Survey[91]

"There is a different and more fundamental cost that is independent of the monetary price. That is the energy cost of exploration and production. So long as oil is used as a source of energy, when the energy cost of recovering a barrel of oil becomes greater than the energy content of the oil, production will cease no matter what the monetary price may be".

4.1 Oil prices

Oil prices have been of particular interest recently, largely as a result of substantial increases .

The average price on a barrel of oil has increased from $20 in 1998 to $109 at the time of writing in September 2008 (Sources BP[92] and EIA[93])

There have been rapid increases in oil prices before. There were seven post-war oil crises up to 2003; in none of these cases was a peak oil scenario reached[94]. All crises involved the Middle East, and all involved the deliberate stoppage of exports by oil-producing countries. The main difference between the high oil prices of 2008 and earlier spikes in oil price is that many commentators believe that global production is largely maximised[95].

This theory bears consideration when examining oil prices and production volumes from Saudi Arabia. Saudi Arabia has long been the world's largest oil producer, and is known as a 'swing' producer because it has been able to produce more oil in order to keep supplies going when production has been disrupted elsewhere. In 2004, the Saudi Minister for Petroleum announced that in light of oil prices exceeding $50:

"Saudi Arabia will... increase the Kingdom's production capacity to 11 million barrels per day by intensifying drilling in production fields". [96]

In 2005, he stated that production capacity would be expanded from the current level of 11 million barrels per day to 12.5

24

million barrels per day by 2009 to meet spare demand and maintain spare capacity of at least 1.5 million barrels per day[97]. However, Saudi Arabia's highest average daily output since 1981 was 9.6 million barrels from April to September 2005 [98], when oil was an average of $58.12 per barrel. Its output from January to April 2008, when oil was an average of $101.74 per barrel, was 9.2 million barrels per day.

In October 2007, the former head of exploration and production at the state-owned Saudi oil company Saudi Aramco commented that production was unlikely to increase further:

"The evidence is that in spite of the increases - very large increases - in oil prices over the last four
years, we haven't been able to match that with increasing capacity. So, essentially, we are on a plateau". [99]

The Chief Economist at the International Energy Agency indicated that:

"According to normal economic theory, and the history of oil, rising prices have two major effects... they reduce demand and they induce oil supplies. Not this time". [100]

Libya's acting Oil Minister Chukri Ghanem recently indicated that OPEC countries cannot increase their production of crude oil, saying *"I believe that we don't have much more to produce".* [101]

One economist considers that demand from fast-developing countries with high levels of foreign exchange reserves means that oil prices are unlikely to reduce as they did after previous oil shocks[102]. The analyst who correctly predicted oil prices breaching $80 and $100 recently predicted that oil will reach an average of $150 per barrel by 2010, and more than $200 per barrel by 2012[103], in a report that "stands to drive North American bicycle sales through the roof"[104]. In May 2008, Goldman Sachs published a document that suggested:

"The possibility of $150-$200 per barrel seems increasingly likely over the next 6-24 months".[105]

4.2 Inflation-Adjusted Oil Prices

Inflation means that the real price of oil historically cannot be compared on a like-for-like basis with the price today. One way of achieving this comparison is to account for inflation

Oil prices in 2008[109] are now higher in real terms than at any time in history. However, Deffeyes notes that correcting oil prices to take account of inflation is a "highly circular exercise" because energy prices are a major cause of inflation[110].

4.3 Oil Demand

More than half (52.6 per cent) of the world's oil reserves (including Canadian oil sands) lie within four countries (Saudi Arabia, Canada, Iran and Iraq).
Excluding Canadian oil sands, the five countries with the largest oil reserves in the world, accounting for approximately 54 per cent of reserves, are Saudi Arabia, Iran, Iraq, Kuwait and the United Arab Emirates. All are members of OPEC.

The USA consumes one quarter of global oil.
Just under half (49.4 per cent) of global oil consumption is accounted for by the six largest consumers (USA, China, Japan, Russia, Germany and India).

The IEA considers that oil demand in 2008 will be 87 million barrels per day on average, which is a substantial reduction on its previous forecasts. [114]

26

4.4 Oil Production, Import and Export

The age of oil fields has an important bearing on oil production rates. For example, future oil production from Angola will probably be 14 billion barrels[115] and future contributions from the North Sea will be similar[116], but production rates are expected to grow rapidly in Angola, while North Sea production is projected to decline[117]. Therefore, the size of oil reserves is not sufficient to know whether or not future production will decline or grow; maturity of fields and addition of new fields must also be considered[118].

Of the top 40 oil-producing countries worldwide[119] (accounting for 97.6 per cent of world production), oil production is declining in 14. Oil production has either reached a plateau or there is no clear trend in a further seven of the top 40 oil-producing countries, and production is increasing in 19.

The vice-president of Lukoil, Russia's largest independent oil company, announced in April 2008 that "*the period of intense oil production [growth] is over*", and that last year's Russian oil production of 10 million barrels per day was the highest he would see "*in his lifetime*".[120]

The UK production peak was 2.68 million barrels per day, in 1999[121]; at this level, the UK was responsible for 4.1 per cent of world production. UK production in 2007 (1.50 million barrels per day) was 44 per cent less than the peak level, and accounted for 2.0 per cent of global production.

Other declining countries are the United States of America, Mexico, Argentina, Colombia, Denmark, Norway, UK, Oman, Syria, Yemen, Egypt, Gabon, Australia and Indonesia. These countries account for 25.3 per cent of production.

Countries that are increasing production are Canada, Brazil, Ecuador, Azerbaijan, Kazakhstan, Russia, Iran, Iraq, Kuwait, Qatar, United Arab Emirates, Angola, Republic of Congo, Libya, Sudan, Brunei, China, Thailand and Vietnam. These countries account for 47.7 per cent of production.

Countries that are on a production plateau or where there is no clear trend are Venezuela, Saudi Arabia, Algeria, Equatorial Guinea, Nigeria, India and Malaysia. These countries account for 24.6 per cent of production.

The former Secretary of State for Trade and Industry, Alastair Darling, confirmed:

"The UK's reserves of oil and gas are declining. While significant amounts still remain in the North Sea, production has hit its peak and is now falling. We will make the most of the reserves we have, but as our economy grows, we will become increasingly dependent on imports in a world where supplies are concentrated in less stable regions"[125].

The UK has been importing oil since 2006.

Exports are calculated as oil production minus oil consumption (although this balance may not hold precisely because of stock draw-down and accumulation, and because some countries import crude oil to refine and export).[126]

Average exports in the years covered (usually 2006) were 40.4 million barrels per day. Exports from OPEC countries, Russia and Mexico (81.4 per cent of all exports) are anticipated to

28

decline by 2.5 million barrels per day by 2012.[127]

Oil Exporting Countries of the World

12 of the top 20 oil-exporting countries in the world are members
of OPEC (Indonesia is the only member that is a net importer of
oil, and it has stated its intention to leave the cartel[128]).
Their daily exports are 24.5 million barrels[129]. The total exports
of non-OPEC countries are 15.9 million barrels[130].

Some of the largest oil exporters have reported very large
increases in oil consumption. Over the past five years, oil
demand in Kuwait and Saudi Arabia has increased by 5 per cent
per year; in Iran, the increase has been 7 per cent per year, and
over the period 2004-07, Venezuela has reported an annual
increase in oil demand of 10 per cent.[131]

The increase in OECD demand was less than 1 per cent per year
over the last five years[132]. The growth in domestic demand from
countries with substantial oil exports is considered likely to
"cannibalize" export capacity, reducing the availability of oil for
importing countries.[133]

This increasing demand is at least in part a result of the subsidies
that many oil-producing countries apply to domestic pricing.
Petrol costs in the order of 7 cents per litre in Venezuela, and
between 13 and 16 cents per litre in Saudi Arabia, Kuwait and
Iran[134]. Since higher oil prices more than compensate for higher
domestic subsidies, reduced demand in these countries is
'unlikely'.[135]

India, Indonesia, Malaysia, Sri Lanka and Taiwan have all
revised their administered prices of fuel,[136] as a result of
reassessing the 'budgetary reality' of maintaining subsidies on

domestic fuel prices,[137]

Such subsidies can lead to outbound smuggling to non-subsidised countries.[138]

China's subsidy of its domestic oil market is calculated as being about \$87 billion per year: more than 2 percent of GDP, and roughly 10 per cent of the government's fiscal revenues.[139] India's subsidy is regarded by the IEA as "financing a massive transfer of wealth to the growing black market".[140]

In June 2008, Malaysia announced it was raising petrol prices by 40 per cent.[141] Prior to the move, the government's fuel subsidies accounted for one third of total spending, and were equivalent to 7 per cent of GDP.

Not all oil production is available on the free market. Importing countries enter into agreements with exporting countries in order to gain preferential access to oil.[142] China's policy of acquiring international petroleum reserves (between June 2005 and June 2006, China invested equity in 0.3 per cent of global proven reserves.[143]) is described as a "mercantilist view of global energy resources", which "does not promote international cooperation in addressing limited supplies of petroleum".[144]

Rising food prices have put biofuel policy under scrutiny[145]. The IEA describes as 'sobering' the scenario of replacing the biofuel component of US and European fuels with mineral oil, at around 1 million barrels per day[146]. The European Parliament Environment Committee has confirmed its opposition to an EU biofuels consumption target for road transport,[147] and there is speculation that the target will be adjusted downwards or removed completely[148].

A Royal Society report published in January 2008 also highlighted the risk that biofuels could fail to deliver significant reductions in transport emissions, while creating harmful environmental and social impacts[149]. Also in January 2008, the

Environmental Audit Committee of the House of Commons requested a moratorium on biofuel targets[150].

5. Consequences of Peak Oil (An Overview)

The Swedish Government has set a new policy target: the creation of the conditions necessary to break Sweden's dependence on oil by 2020. And there is, indeed, an increased sense of urgency.

"If we prepare now, the transition to a sustainable energy system can be smooth and cost-efficient.
If we wait until we are forced by circumstances, the transition may be costly and disruptive. No country can escape from this transition; to act sooner or act later are the only options... It is already a major competitive advantage for Sweden's industry and the economy that, by international standards, Sweden is not so dependent on oil".

Mona Sahlin, Swedish Minister for Sustainable Development, 9 May 2006[158]

To an extent, the consequences of peak oil depend on the preparation that countries make to buffer the impacts. For example, early mechanisms to decrease oil consumption and encourage substitution would provide better preparation than simply waiting for the oil price to suppress demand. Hirsch considers *"aggressive risk management"* to be *essential to address the challenges arising from peak oil*[159].

Sweden is the only country that has a government commitment to breaking its dependence on oil.

The Commission on Oil Independence is tasked with ensuring that Sweden is *"free of dependence on fossil fuels for transport and heating by 2020"*.[160]

Swenson has examined some scenarios for energy demand that could be followed as a consequence of peak oil. They are described in the Swenson curve.

Each section of the post-peak curve describes the adaptation mechanisms that could be used to reduce the impact of reduced oil supply:[161]

Conservation is described as enjoying a similar lifestyle, but accomplished with more energy-efficient artifacts, such as more energy-efficient appliances (although conservation is described elsewhere as the painful rationing that high prices will force on ordinary people).[162]

Examples of lifestyle change include people telecommuting instead of commuting, living a more community-based lifestyle, or living closer to work.

Substitution is effected by using other energy sources to accomplish the same objectives (solar power, walking or cycling instead of driving).

Deprivation means doing without. It includes scenarios such as mass hunger and starvation, and war over resources.

The implication is that since deprivation is more unpleasant than the alternative adaptation measures, action should be taken to encourage the three other measures before the peak is reached[163], particularly since 95 per cent of the energy used in the transport sector comes from oil[164]. Campbell notes that those countries that do plan and prepare for peak oil "will clearly have great advantage over those that simply react to the crisis when it hits them".[165]

The former US Energy Secretary noted that a major economic shock − and political unrest − would be the result of failing to prepare for the peak of conventional oil production[166].

Ayres comments that:

"The economy is utterly dependent on petroleum... and I think it is highly likely that when oil production peaks, so will the world economy. When petroleum gets more expensive everything that depends on it gets more expensive, and I cannot see how growth could really continue with much more expensive energy".[167]

The first manifestation of peak oil is likely to be an increase in oil price, as demand exceeds supply, and competition for resources takes place.[168] A "supply-side crunch in the period to 2015" is regarded by the IEA as a possibility, which would involve "an abrupt escalation of oil prices".[169]

At a certain level, increasing prices would force demand reductions[170], which would act to stabilise, and possibly temporarily reduce, prices.[171] However, rising energy prices increase business costs, so a rise in the real price of oil "has to be accommodated by a fall in the real wage".[172] Professor Oswald suggests that oil price rises take about 18 months to feed through noticeably to the real economy.[173]   One report suggests that the decade following the peak will feature dramatic increases in inflation, long-term recession, high unemployment and declining living standards.[174]

A report commissioned by the United States Government noted that mitigation of peak oil would require a decade of "intense, expensive effort", and that intervention by governments will be required, because the economic and social implications of peak oil would otherwise be "chaotic."[175]

The report goes on to note that prudent risk management requires the implementation of mitigation "well before peaking"[176]. The problems of peak oil will be "especially serious" for developed countries, while problems in developing countries have the potential to be "much worse"[177].

In May 2008, the Ernst and Young Item Club warned that if oil prices remain at their current levels, it would have to cut its economic growth forecast for the UK for next year to 1.3 per cent; inflation would also be higher than 3 per cent for the next three years.[178]

In April 2008, the Executive Director of the International Energy Agency claimed that an oil price of $118 per barrel would raise concerns about the global economy's ability to avoid entering a recession.[179]

A document written by the former Chief Economist at the US President's Council of Economic Advisers suggests the following likely responses to an oil price of $120 per barrel:[180]

Disruptions to normal economic activity, including factory

shutdowns, reduced long-distance travel, and layoffs in key

industries which spread to the rest of the economy.
Increased taxation will be necessary to cover the increased costs
faced at all levels of government.
Household energy bills (including automobile transport costs)
will roughly double to 10-15 per cent of family budgets.

A sharp reduction in revenues for travel, tourism, and automobile
businesses.

Inflation will rise from 2-3 per cent to 6-8 per cent, requiring
decisions to be made over interest rates that may make economic
growth rates 'tumble'.

A 25 per cent decline in global stock market valuations.

A cut in world GDP by up to 3.6 percentage points.

The onset of global recession.

The situation up to and beyond peak oil is described by Campbell
as likely to be:

*"a period of recurring price surges, recessions, international
tensions, and growing conflicts for access to critical oil supplies,
as the indigenous energy supply situation in the United States
and Europe deteriorates"*[181],

and peak oil supply constraints will curb demand by higher
prices, so that *"the historical pattern of economic growth has to
end"*[182]. Increasing oil prices will feed through into increasing
food prices.[183] Rising prices of food and energy will hit poor
people hardest.[184]

An analyst at the Ernst and Young Item Club suggested that if oil reached $200 a barrel permanently, the Governor of the Bank of England would suffer from "writer's cramp" with the number of letters he would have to write to the Chancellor explaining why the UK economy had breached the 2 per cent inflation target.[185]

In the most optimistic scenario, the modelling used for the Energy Information Administration's report International Energy Outlook 2007 suggests that long-term projections for economic growth will not be affected "substantially" by oil price, because by 2030 the global economy will have adjusted to different oil prices[186]. However, in the shorter term, following the 'high world oil price' scenario:

"As higher oil prices feed through the economy and reduce purchasing power, real aggregate expenditures on goods and services decline... unemployment increases, energy-intensive capital stock begins to become obsolete, and real GDP is lower. In oil-importing countries that also have
major oil producing sectors, like the United [Kingdom], higher oil prices increase the flow of economic resources into oil production activities.
At the same time, national expenditures on petroleum imports increase, with negative repercussions for real GDP. Countries wholly dependent on oil imports, like Japan, are forced to spend more for their energy purchases... In the medium term, increases in unemployment lead to downward adjustments in wages and prices. In developed countries, central banks react by lowering key policy rates, thus boosting interest-sensitive aggregate demand. After 2015, the rebound effects of lower employment costs, lower prices, and lower interest rates outweigh the contractionary effects of higher oil prices, leading to stronger real GDP growth and lower inflation".[187]

36

The more immediate problem, however, is described by Deffeyes:

"Our transportation system is almost totally driven by products from oil. As we learned in the late 1970s, an oil shortage can ripple through the economy, lowering our standard of living. My concern is not about our long-term adaptation to a world beyond oil. Through our inattention, we have wasted the years that we might have used to prepare for lessened oil supplies. The next ten years are critical.
It's going to be on-the-job training. Learn while doing: not always the most orderly way of adapting".[188]

In conclusion we can summarise the following:

-Oil supply is about to fall.
-Oil demand is increasing.
-Mass-market alternatives to oil are not being found fast enough.

Given these combination of factors, we are potentially facing a huge economic crisis.

Chapter 3
The Complacency Surrounding Peak Oil

"A man is his own easiest dupe, for what he wishes to be true he generally believes to be true."
-Demosthenes c.383-322 BC

"Anybody who believes exponential growth can go on forever in a finite world is either a madman or an economist."
-Kenneth Boulding

Predicting the end of any natural resources has the attendant problem of dealing with past erroneous predictions. In particular, the predictions of the pre 1980s often went spectacularly wrong in light of the next two decades. The aftermath of these early doomsday pronouncements has led to an anesthetizing of the public, even the informed public, against further pronouncements of this ilk. The environmental movement has cried wolf once too often and now they are not going to be believed. But let's investigate this premise more carefully.

The predictions, with regards to non renewable resources, have been one area which has been particularly subject to error. The Club of Rome reports (Meadows et al 1972)[1], which first came out in 1972, suggested that the world would be limited by two fundamental restrictions, either lack of resources or by an increase in pollution. The long term predictions were severe and suggested drastic world population declines sometime in the middle of the twenty first century, this century. Unfortunately the report focused more on minerals rather than energy and on the pollutant effects of chemicals other than CO_2. The anthropogenic greenhouse effect was just coming on to the world's radar at this time and was not yet quite thought to be a serious threat.

The Earth's mineral resources exist in various proportions in the environment with the easily mined parts having concentrations which allow relatively convenient extraction using modest amounts of energy. As the better ores are mined the industry is forced to go to lower and lower grade ores; ores that generally need greater and greater amounts of energy to facilitate the removal of the mineral of concern. As long as the economics are right, the move to poorer ores can be accommodated and for all intensive purposes the reserves of most minerals are in fact pretty much inexhaustible, in the sense that they will never be actually exhausted. Most elements for instance are present in sea water, at least at the parts per trillion levels, and there is a lot of sea water. So as long as extraction economics are viable, and this usually means the energy needed is available and sufficiently cheap, the economy will not be limited by lack of minerals in the foreseeable future. This reasoning, and the primacy of energy resources, was not immediately apparent in the earlier doomsday reports and has led to confusion and a severe loss of face by people wishing to protect resources for future generations.

Crying wolf:

The significance of the problem that the earlier reports have caused to any rational investigation of the current world situation cannot be overemphasized. One such incident which gained notoriety was a bet between an economist from the University of Maryland, Julian Simon [2] and the (doomsday) environmentalist Paul Ehrlich [3]. Ehrlich was intent on proving that the world would soon run out of resources and a population collapse would ensue. Simon knew the market better and was convinced that market economics would operate and that increased consumption would result in a reduction in costs and (or) substitution for the resource whose price went too high due to shortages. A bet was arranged with Ehrlich choosing 5 metals (copper, chrome, nickel,

tin, and tungsten) which were to be purchased for a total of US$1,000. The idea was that if the price of the minerals went up, as Ehrlich expected, in a ten year period, then Simon would have to pay up the difference in total cost. If, as Simon expected, the price went down then Ehrlich would have to pay up.
The bet was resolved in 1990 and went clearly in the favor of market economics and so Ehrlich paid Simon a cheque to the value of US $576.07. [4]

Failures in prediction by environmentalists, as illustrated in this episode, pretty much dented further efforts to awake the world to a crisis based on shortages of resources. Pollution fared somewhat better to the extent that global warming reached the world stage in the series of negotiations undertaken culminating in the Kyoto agreement. Local level pollution, however, took more of a back seat, in that the more obvious problems were predominantly solved using improved technology and adherence to stricter guidelines, further diminishing environmentalist claims that disaster was just around the corner. In general the global economy has boomed over the last two decades with some of the larger third world nations, including China, aspiring to approach first world standards.

Physical reality versus economic reality:

So what is the problem? The problem is that the world economic picture, painted by economists on a background of free trade and globalisation, has failed to take into account the physical reality of energy in general and fossil fuel resources in particular; especially in terms of the state of the world's crude oil supplies. An examination of oil supplies will lead us to the conclusion that we have essentially been led up the garden path by a system of unrestricted marketing and growth dominated economics.

That there has been very little serious outcry or realisation of the

situation, at least from the developed world, might be attributed
to the fact that in general we are comfortable and have never had
it so good. And that scientists have further abdicated the
operation of the global human situation to economists, due to the
supposedly self evident success of this approach and the fact that
contrary environmentalist meddling could be dismissed in the
light of their obvious failures in the past.

Peak Oil Questions and Concise Answers

Q1) Is anyone in the oil industry confident of being able to meet future world energy demands?

The energy industry has quietly acknowledged the seriousness of the situation. For instance, in an article recently posted on the Exxon-Mobil Exploration homepage, company president Jon Thompson stated:

"By 2015, we will need to find, develop and produce a volume of new oil and gas that is equal to eight out of every barrels being produced today. In addition, the cost associated with providing this additional oil and gas is expected to be considerably more than what the industry is now spending.
Equally daunting is the fact that many of the most promising prospects are far from major markets — some in regions that lack even basic infrastructure. Others are in extreme climates, such as the Arctic, that present extraordinary technical challenges". [viii]

If Mr. Thompson is that frank in an article posted on the Exxon-Mobil Webpage, one wonders what he says behind closed doors. The Saudis are no less frank than Mr. Thompson when discussing the imminent end of the oil age. They have a saying that goes, "My father rode a camel. I drive a car. My son flies a jet airplane. His son will ride a camel."

Q2) Why should I be particulary concerned? Rising petrol (gasoline) prices are not the end of the world. We can use more efficient cars or drive less.

Almost every current human endeavor — from transportation, to manufacturing, to electricity, to plastics, and especially food and water production — is inextricably intertwined with oil and natural gas supplies.

A. Oil and Food Production

In the US, approximately 10 calories of fossil fuels are required to produce 1 calorie of food. [10] If packaging and shipping are factored into the equation, that ratio is raised considerably. This disparity is made possible by an abundance of cheap oil. Most pesticides are petroleum-
(oil) based, and all commercial fertilizers are ammonia- based. Ammonia is produced from natural gas, a fossil fuel subject to a depletion profile similar to that of oil. Oil has allowed for farming implements such as tractors, food storage systems such as refrigerators, and food transport systems such as trucks. Oil-based agriculture is primarily responsible for the world's population exploding from 1 billion at the middle of the 19th century to 6.3 billion at the turn of the 21st. As oil production went up, so did food production. As food production went up, so did the population. As the population went up, the demand for food went up, which increased the demand for oil.

Within a few years of Peak Oil occurring, the price of food will skyrocket as the cost of producing, storing, transporting, and packaging it will soar.

B. Oil and Water Supply

Oil is also needed to deliver almost all of our fresh water. Oil is used to construct and maintain aqueducts, dams, sewers, wells, as well as to pump the water that comes out of our taps. As with food, the cost of fresh water will soar as the cost of oil soars.

C. Oil and Health Care

Oil is also largely responsible for the advances in medicine that have been made in the last 150 years. Oil allowed for the mass production of pharmaceutical drugs, surgical equipment and the development of health care infrastructure.

D. Oil and Everything Else

Oil is also required for nearly every consumer item, sewage disposal, rubbish disposal, street/park maintenance, police, fire services, and national defense. Thus, the aftermath of Peak Oil will extend far beyond how much you will pay for petrol (gasoline). Simply stated, you can expect: economic collapse, war, amd widespread starvation.

Q3. People predict disaster situations all the time which are rarely true.
Where are you getting this information from? Who else is talking about Peak Oil? What type of backgrounds do they have? How do I know they are credible?

Peak oil is not your usual conspiracy theory created by paranoid anti-establishment types or militant anti-capitalists.
It is a self-evident truth that oil will run out at some point. As supply dwindles our world economy will suffer.
The only argument lies in when exactly it will happen and how bad the effects will be.
This book is supported by an analysis of hard facts reported by highly respected sources. Some of the more notable sources are described below. As you will see, this is not the usual "world is nigh" crowd.

A. Petroconsultants Pty Ltd

In 1995, Petroconsultants Pty Ltd. one of the largest and most respected oil industry analysis and consulting firms, released a document called, "World Oil Supply 1930-2050." This report, which was written for oil industry insiders and cost a huge $32,000 per copy, predicted that global oil production will peak around the year 2000 and decline by 25% by 2025. [23]

B. Dr. Colin J. Campbell, Former Exploration Geologist for Texaco and Chief Geologist for Ecuador

In a February 2002 report Dr. Campbell explained:

"Peak Oil is a turning point for mankind. The economic prosperity of the 20th Century was driven by cheap, oil-based energy. Everyone had the equivalent of several unpaid and unfed slaves to do his work for him, but now these slaves are

*getting old and won't work much longer. We have an urgent need
to find how to live without them."* [24]

C. Dr. David Goodstein, Professor of Physics and Vice Provost of Cal Tech University

In his book, *Out of Gas: The End of Oil*, Dr. Goodstein argues
forcefully that the worldwide production of oil will peak soon,
possibly within this decade. That will be followed by declining
availability of fossil fuels that could plunge the world into global
conflicts as nations struggle to capture their piece of a shrinking
pie. In a recent interview with ABC news, Dr. Goodstein had this
to say about Peak Oil:

*"Best case? The worldwide disruptions that follow the peak serve
as a global wake-up call. A methane-based economy is successful
in bridging the gap temporarily while nuclear power plants are
built and the infrastructure for other alternative fuels is put in
place. The world watches anxiously as each new Hubbert's peak
estimate for uranium and oil shale makes front-page news."*

*Worst case? After the peak, all efforts to produce, distribute, and
consume alternative fuels fast enough to fill the gap between
falling supplies and rising demand fail.
Runaway inflation and worldwide depression leave many billions
of people with no alternative but to burn coal in vast quantities
for warmth, cooking, and primitive industry. The change in the
greenhouse effect that results eventually tips Earth's climate into
a new state hostile to life. End of story".* [25]

D. Matthew Simmons, Former Energy Advisor to George W. Bush and considered by many peers to be the leading world authority on oil production.

In an August 2005 interview with FromTheWilderness.com

46

publisher Michael Ruppert, Mr. Simmons was asked if it was
time for Peak Oil to become part of the public policy debate.

He responded:

*"It is past time. As I have said, the experts and politicians have
no Plan B to fall back on.
If energy peaks, particularly while 5 of the world's 6.5 billion
people have little or no use of modern energy, it will be a
tremendous jolt to our economic well-being and to our health —
greater than anyone could ever imagine.*

When asked if there is a solution, Simmons responded:

*I don't think there is one. The solution is to pray. Under the best
of circumstances, if all prayers are answered there will be no
crisis for maybe five years. After that it's a certainty".*[26]

Dick Cheney, Vice President of the USA

In late 1999, Dick Cheney stated,

*"By some estimates, there will be an average of two percent
annual growth in global oil demand over the years ahead, along
with, conservatively, a three-percent natural decline in
production from existing reserves. That means that by 2010 we
will need on the order of an additional 50 million barrels a
day."*[27]

This is equivalent to six times the amount of oil produced per day
by Saudi Arabia, the world's leading oil producer.

A report commissioned by Cheney and released in April 2001 was no less rosy, *"The most significant difference between now and a decade ago is the extraordinarily rapid erosion of spare capacities at critical segments of energy chains. Today, shortfalls appear to be endemic. Among the most extraordinary of these losses of spare capacity is in the oil arena."* [28]

Q4) Is it possible that we have already hit Peak Oil and are now in the first stages of the Oil Crash?

Yes, although we cannot be sure. A reasonable amount of evidence exists that we have already hit peak supply.

A. Declining Oil Production

In May 2003, at the Paris Peak Oil Conference, Princeton Professor Kenneth Deffeyes, author of *Hubbert's Peak: The Impending World Oil Shortage,* explained that Peak Oil actually arrived in 2000 by noting that production has actually been declining since that time. [29]

B. Drastically Revised Estimates of Oil & Natural Gas Reserves

In October 2003, *CNN International* reported that a research team from Sweden's University of Uppsala has discovered worldwide oil reserves are as much as 80% less than previously thought, that worldwide oil production will peak within the next 10 years, and once production peaks, gas prices will reach disastrous levels. [30]

In January 2004, shares of major oil companies fell after Royal Dutch/Shell Group shocked investors by slashing its "proven" reserves 20 percent, raising concerns others may also have improperly booked reserves. [31]

A month later, energy company El Paso Corporation announced it had cut its proven natural gas reserves estimate by 41 percent. [32]

48

C. Generally High Oil Prices

On June 28 2008, the price of oil hit $142.99 a barrel, the highest price ever. Although it has since come down because of a slow down, in the world economy. The long-term average will generally increase and will only come down during periods of recession. The days of $20 dollar barrels are long gone.

D. Reduced Food Production

World grain production has dropped every year since 1997. [34] World wheat production has dropped every year since 1998. Food price hikes in China could be the sign of a coming world food crisis brought on by global warming and increasingly scarce water supplies among major grain producers. [35]

E. Conclusion

If you were to look at any one of these pieces of evidence in isolation, it would not tell you much about the situation the world is in. However, when you look at all of them together in the context of Peak Oil, the fact that we are already crashing becomes obvious.

**Q5) What about the oil in the Arctic National Wildlife
Preserve (ANWR)?
If the environmentalists got out of the way, couldn't we just
drill for oil there?**

At current rates of oil consumption, the ANWR contains enough
oil to power the US for only six months. [39]
The fact that it is being touted as a "huge" source of oil
underscores how serious our problem really is.

**Q6) What about the oil under the Caspian Sea? I heard there
was a massive amount of oil underneath it.**

As recently as September 2001, the Caspian Sea was thought to
be the oil find of the century. By December 2002, however, just
after US troops took Afghanistan, British Petroleum announced
disappointing Caspian drilling results. The "oil find of the
century" was little more than a drop in the ocean. Instead of
earlier predictions of oil reserves above 200 billion barrels, the
US State Department announced, *"Caspian oil represents 4% of
world reserves. It will never dominate the world's markets."* [40]

Furthermore, the area has the potential for wars and disruptions
that could make the Persian Gulf look tame by comparison.
Unstable countries surround the Caspian, including Russia,
Kazakhstan, Turkmenistan, Uzbekistan, Iran, and Azerbaijan.
Proposed pipelines to carry the oil run through hotspots such as
Afghanistan, Pakistan, Turkey, China, Russia, Ukraine, Bulgaria,
and Kyrgyzstan. Meanwhile, the region is isolated and
unforgiving, so the expenses associated with drilling would be
enormous. [41]
Despite these monumental obstacles, oil is becoming so scarce
that even the disappointingly modest amounts located in the
Caspian Sea will remain extremely important from a geopolitical
standpoint.

**7. What about so-called "non-conventional" sources of oil?
Doesn't Canada have an enormous amount of this type of oil?**

So called "non-conventional" oil, such as the oil sands found in
Canada and Venezuela, is incapable of replacing conventional oil
for the following reasons:

1. Non-conventional oil has a very poor energy/profit ratio, and
is extremely difficult to produce. It takes about two barrels of oil
in energy investment to produce three barrels of oil. [16]
The cost of Canadian non-conventional oil projects is so high
that in May 2003, the oil industry publication *Rigzone* suggested,

*"President Bush, known for his religious faith, should be praying
nightly that Petro- Canada and other oil sands players find ways
to cut their costs and boost US energy security."* [43]

2. The environmental costs are horrendous and the process uses a
tremendous amount of fresh water and also natural gas, both of
which are in limited supply.

3. Although non-conventional oil is quite abundant, its rate of
extraction is far too slow to meet the huge global energy demand.
Dr. Colin Campbell estimates that combined Canadian and
Venezuelan output of non-conventional oil will be 3.6 million
barrels per day (mpd) in 2012, and 4.6 mbd in 2020. These are
minor sources, given today's consumption of 84 mbd, which is
expected to increase to 120 mbd by 2020. [44]

Q8) Is it possible that there is still major oil fields left to be discovered?

It is very unlikely. All available evidence indicates that we have already located most of the world's oil reserves:

A. World's Largest Oil Fields Are All Over 40 Years Old

According to a recent report from the Colorado School of Mines entitled *The World's Giant Oilfields,* the world's 120 largest oilfields produce almost 50% of the world's crude oil supply. The fourteen largest account for over 20%. The average age of these 14 largest fields is 43.5 years." [48]

The reserves in the world's super-giant and giant oilfields are dwindling at an average rate of 4-6 percent a year. [49] The study concludes that "*most of the world's true giants were found decades ago.*"

B. Very Few New Oil Discoveries Made in Recent Years

Over the past 20 years, despite investment of hundreds of billions of dollars by major oil companies, results have been alarmingly disappointing. Most of the world has now been digitally "X-rayed" using satellites and seismic data in the process of locating 41,000 oil fields. Over 641,000 exploratory wells have been drilled, and virtually all fields which show any promise are well-known and factored into the one-trillion barrel estimate the oil industry uses for world oil reserves. [50]

A recent study published in *Petroleum Review* suggests that production might not be able to keep up with demand by 2008. [51] The study is a survey of oil "mega projects," which are projects containing more than 500 million barrels — or the amount of oil the world consumes in 5-10 days. The discovery rate for mega

projects has dwindled to almost nothing. In 2000, 16 mega projects were discovered.[52]
In 2001 there were only 8 new discoveries, and in 2002 there were only 3 such discoveries.[53]

It's not just "mega projects" that are no longer being found. Very few oil projects of any size are being discovered these days. Between 2001 and 2003, oil companies discovered less than half the reserves found between 1998 and 2000.[54] About 80% of the oil produced today flows from fields that were found more than 30 years ago, and the great majority of them are declining.[55]

C. US Mainland as an Example

If you need any more convincing that there are no (or at least very few) large oilfields left to be discovered, consider the example of the US mainland. In the US they have been searching for and extracting oil for longer than anyone, and have had more financial and technological muscle than anyone. If anyone could turn around oil declines, it would be us. US oil production peaked in 1970 and has consistently fallen since then. Despite maximum financial incentives, the finest technology in the world and a complete openness to innovation, the US has been unable to slow, never mind reverse, this 2%-per-year production decline.[56]
Thirty-five years of money and research has neither slowed nor reversed the decline. Why then should the world fare any better? The truth is the world won't fare any better.[57]

D. Conclusion
Matthew Simmons has stated succinctly,

"All the big deposits have been found and exploited. There aren't going to be any dramatic new discoveries, and the discovery trends have made this abundantly clear." [58]

On a similar note, according to Dr. David Goodstein, *"Better to believe in the Tooth Fairy than the possibility of any more large oil discoveries."* [59]

Q9) I know Simmons, Goodstein and others like them are experts with impeccable credentials, but even experts turn out to be wrong sometimes. What if there is actually a huge amount of oil just waiting to be discovered? Wouldn't that make a difference?

Not significantly. Given huge increases in consumption rates, even if we begin finding staggeringly massive amounts of oil tomorrow, the crisis would only be delayed by a few years. As University of Colorado physicist Albert Bartlett explains, given exponential consumption growth, *"a doubling of the remaining resource results in only a small increase"* in its life expectancy. Bartlett's recent mathematical analysis of oil production calculated that even if the estimated total ultimate oil supply were doubled from 2,000 billion barrels to 4,000 billion barrels, it would only delay the peak by 25 years [60]

This means that if even the current amount we think we have left is tripled from 1,000 billion barrels to 3,000 billion barrels, we are still within one generation of serious problems.

Q10) Wont technological advances mean we have more efficient ways to use the oil that is left?

Absolutely, but increased efficiency won't be proportional to increased demand. World oil demand is set to increase by 50% over the next 20 years. It is extremely unlikely that effeciency will increase anywhere near 50% in that time period.
We have become more efficient over time already and yet have continued to use much more oil. That trend must be reversed if we are to effectively cope with this crisis. In a free-market economy, however, that is unlikely, since by the time market forces make alternatives (assuming there any) for oil cost effective, the crisis will have already gathered massive momentum. We shall discuss this point later.

Q11) I heard that some scientist has a theory that fossil fuels actually renew themselves. If that's true, wouldn't it cast doubt on the validity of Peak Oil?

The scientist you speak of is a man by the name of Dr. Thomas Gold. In his 1999 book, *The Deep Hot Biosphere,* he proposes a theory that oil comes from deep in the Earth's crust, left over from some primordial event in the formation of the Earth, when hydrocarbons were formed. [61]

If his theory were true, it would mean that fossil fuels are actually renewable resources.

Unfortunately, his theory has been proven to be false, time and time again. As Steve Drury, who reviewed Gold's book for *Geological Magazine,* puts it, *"Any Earth scientist will take a perverse delight in reading the book, because it is entertaining stuff but even a beginner will see the gaping holes where Gold has deftly avoided the vast bulk of mundane evidence regarding our planet's hydrocarbons."* [62]

When asked about the validity of theories such as Gold's, Dr. Colin Campbell responded:

Oil sometimes does occur in fractured or weathered crystalline rocks, which may have led people to accept this theory, but in all cases there is an easy explanation of lateral migration from normal sources. Isotopic evidence provides a clear link to the organic origins. No one in the industry gives the slightest credence to these theories: after drilling for 150 years they know a bit about it. Another misleading idea is about oilfields being refilled. Some are, but the oil simply is leaking in from a deeper accumulation. [63]

Finally, the deep-earth hypothesis has a fatal flaw: If oil were, indeed, formed under intense heat and pressure in the center of the Earth, it would tend to disintegrate as it rose from the regions of high temperature and pressure to the benign, cooler, low-pressure world closer to the Earth's surface. [64]

Q11. If this were all true, wouldn't you see petrol (gasoline) stations closing on a large scale ?

Since September 2003 Chevron-Texaco has disposed of 1200 petrol (gas) stations in the United States; 900 in Asia and Africa; retail and refining operations in Europe, South America, Australia and the Middle East; and exploration and production holdings in North America, the North Sea and Papua. These are the actions of a company planning a profitable decline strategy, which indeed would be realistic given the pending peak and decline of world production. The action delivers a broader message. [65]

Q12. Didn't the Club of Rome make this exact same prediction back in the 1970s?

In 1972, the Club of Rome (COR) shocked the world with a study titled *The Limits to Growth*, which concluded that:

1. If the population continued to grow and industrialize as it had been, society would run out of renewable resources by the year 2072. Mass depopulation would ensue.

2. Even if the supply of resources was magically doubled, a collapse would occur as a result of pollution. [66]

Often, whenever somebody makes an "end of the world"-type prediction, they are derided as a "Club of Romer." This is extremely unfortunate, as it appears the COR turned out to be correct.

Matthew Simmons, who stated in 2000,
"In hindsight, The COR turned out to be right. We simply wasted 30 important years by ignoring this work." [67]

On a similar note, Richard C. Duncan of the Institute on Energy recently noted:

"Forecasts of the imminent depletion of oil are as old as the industry itself, and that has not changed. What has changed is the growing amount of historical oil data now available to test forecasts. Of the 44 significant oil-producing nations, at least 24 are clearly past their peak of production." [68]

Thus the discussion about global oil production is now less about predicting the future as it is about recording history as it happens.

Q13) We had oil problems back in the 1970s. How is this any different?

The oil shortages of the 1970s were the results of political events. The coming oil shortage is the result of geological reality.

As far as the US oil supply was concerned, in the 1970s there were other 'swing' oil producers like Venezuela who could step in to fill the supply gap. Once worldwide oil production peaks, there won't be any swing producers to fill in the gap. In a recent issue of the Association for the Study of Peak Oil Newsletter, Dr. Colin Campbell explained how the coming global oil shortage will be different from previous oil shortages:

"To date, when an individual country, province or region reached this point (production decline), it simply bought from elsewhere. But what will happen when decline is a global phenomenon? Initially it will be denied. There will be much lying and obfuscation. Then, prices will rise and demand will fall. The rich will outbid the poor for available supplies. The system will initially appear to rebalance.
The dash for gas will become more frenzied. People will realise nuclear power stations take up to ten years to build. People will also realise wind, waves, solar and other renewables are all pretty marginal and take a lot of energy to construct. There will be a dash for more fuel-efficient vehicles and equipment. The poor will not be able to afford the investment or the fuel.
Exploration and exploitation of oil and gas will become completely frenzied. More and more countries will decide to reserve oil and gas supplies for their own people. Air quality will be ignored as coal production and consumption expand once more. Once the decline really gets under way, liquids production will fall relentlessly by 5% per year. Energy prices will rise remorselessly. Inflation will become endemic. Resource conflicts will break out." [69]

Q14) How does all this tie in with Global Climate Change?

Fossil fuel consumption and global climate change are intricately tied to each other.
According to data analyzed by the Global Commons Institute there has been a near 100% correlation between world Gross Domestic Product growth and the emission of greenhouse gases from the consumption of hydrocarbon energy.

It now appears that we will have to deal with the implications of Peak Oil at the same time we finally have to pay the piper in regards to global climate change. According to a report recently released by the Pentagon, climate change could bring the planet to the edge of anarchy as countries develop a nuclear threat to defend and secure dwindling food, water and energy supplies.

The report indicates that by 2020 "catastrophic" shortages of water and energy supply will become increasingly harder to overcome, plunging the planet into war. The report concludes that, *"An imminent scenario of catastrophic climate change is plausible and would challenge US national security in ways that should be considered immediately."* [71]

Q15) There is no way humanity will suffer as result of Peak Oil.
Humans are resourceful and we will find a solution if we have to.

Perhaps, but as it currently stands right now, this premise is only based on optimism, blind faith in technology or market forces. Current geological data and our current technological capabilities indicate there will be problems.

Although there are alternatives, we are not developing them anywhere near fast enough. The vast majority of the world's population are not even aware of the problem, let alone preparing for it and attempting to avert it.

Even most polticians are not even aware of the problem and only international poltical cooperation and intervention can avert a crisis.

What can we do about it?

This book has endeavoured, so far, to prove the case for one big argument. First, there is plenty of oil and gas left, but not enough to feel growing global energy demand for much longer. The oil topping point, otherwise known as the peak of producton, will be reached in the 2008-2015 window or it may have already been reached.
When the market realizes this, severe economic trauma will ensue.
Beyond this premise, I now make five arguments, one leads on from the other:

1. It will be possible ro replace oil, gas and coal completely with a plentiful supply of renewable energy. (We can think of this in a very real sense as "endless power", and we will get to that promised land one day. When we do, we'll wonder why it ever took us so long.)

2. However - a very big however - the shortfall berween current expectation of oil supply and actual availability will be such that neither gas, nor renewables, nor liquids from gas and coal, nor nuclear, nor any combination thereof, will be able to plug the gap in time to head off the economic trauma resulting from the oil topping point. (Stated another way we've left it too late. Many people from all socio-economic strata will suffer.)

3.Renewable energy and fuel use, alongside energy efficiency will increasingly substitute for oil and gas, growing explosively whatever happens.

4.However - another very big however - amid the ruins of the old energy modus operandi many will try to turn to coal.
This means that the extent to which renewable energy grows

explosively instead of coal expansion, rather than alongside it, will determine whether economies and ecosystems can survive the global warming threat

5. There is much that people can do to influence the outcome of this struggle to increase renewables' production faster than coal, hence to ameliorate the worst excesses of the global energy crisis, and to create a better society in the process.

62

1. It will be possible to replace oil, gas and coal completely with a plentiful supply of renewable energy and faster than most people think.

Shell employs roomfuls of clever people just to think about the future. They are called scenario planners. In their 2001 book of scenarios, Shell's planners mention that renewable energy holds the potential to power a future world populated with 10 billion people, and do so with ease. The needs of the 10 billion can be met even in the unlikely and undesirable event that all of them use energry at levels well above the average per-capita consumption today in the EU.[250]

The Shell futurists mention this almost in passing, in the caption of a diagram showing the continent-by-continent potential for individual renewable-energy technologies to contribute to such a power-rich furure.
Working for an oil and gas giant as they do, it is perhaps no surprise that they fail to explore this scenario further, Let *us* consider it.

In such a discussion, it is important to emphasize at the outset what might be callecl the "big portfolio" approach to the retreat from fossil fuels. No one technology would be needed to produce all of a nation's energy demand, or even come close. We can mix and match among the family of renewable-energy technologies, because it is a family with a lot of brothers and sisters. We'll have a look at the family in a moment, but we will also need to consider two vital additional components of the story: renewable-energy storage and energy efticiency.

As fossil-fuel die hards are fond of saying, there is not much point in having energy from the elements if it can't be stored. "What happens when the wind doesn't blow?" "What happens when the sun doesn't shine?"

This is where fuel cells, hydrogen and batteries enter the equation. Similarly if the world's routine energy demand profligacy can be deconstructed with efficiency measures, the mountain to climb without oil, gas and coal can be made a lot less steep.

In what follows I am going to consider renewables, storage and efficiency separately though there will be profound overlaps between the three when the world is forced by the oil topping point to attack its energy-supply problems seriously.
Renewables are a big family of options
The first thing you notice about the Shell diagram showing renewable supply meeting the energy demands of 10 billion people wasting energy at the level your average wasteful European does today, is that solar power is the biggest potential contributor to such a hypothetical future.

At one level, this is not perhaps surprising. Enough light falls on the surface of the planet each day to power human society many thousands of times over. Using solar photovoltaic (PV) cells the world's current energy demand - all forms of energy use including transport - could be met using a tiny fraction of the planet's land surface. [251]

For example, the total electricity-generating capacity if all the existing power stations in the world today, of all types, could be created by coverinq an area of the Sahara desert some 600 square kilometres with solar PV[252]

Even in the cloudy UK, more electricity than the nation currently uses could be generated by puttrng PV roof tiles on all suitable roofs[253]

Solar thermal technology, which can be used for both heating and electricity production, holds no less potential.

This member of the family uses collectors to absorb heat from sunlight and then to heat liquid. The heat can then be released from the liquid into a storage tank for use in a building. Where this technique is used for electricity generation, devices that concentrate sunlight, such as mirrored curved reflectors, heat liquid to very high temperatures, creating steam that drives turbines.

"Solar farm" power plants of both solar PV and solar thermal collectors exist today only in small numbers, but will be a colmmon sight - especiallv on otherwise useless scrub land in the sunbelt - once the solar revolution takes off.

Wind power also plays a huge role. America could provide all the electricity it uses today from the wind-power potential of just three states: Texas, North Dakota and Kansas.[254]

Europe's electricity demand could be met using offshore wind farms alone[255]

In the UK's case only a tiny fraction of suitable offshore areas would be needed to meet the nation's total current demand.

The potential of solar and wind extend beyond electricity and heating to transportation. Sufficient hydrogen to fuel every highway vehicle in the United States could be generated, for example, with the wind potential of two states alone, North and South Dakota[256]

As well as solar and wind, our renewable-energy options include marine energy, hydropower, biomass, combined heat and power heat pumps and biofuels.

Let me quickly summarize these technologies and their potential, in turn, before moving on to the close relatives of the family-storage and efficiency.

Both tides and waves can be used to generate marine power. To tap the energy of tides, gates and turbines are installed along a dam or barrage across an estuary or bay.

When the height of the water builds up to the right level on either side of the barrage - on both incoming and outgoing tides - the gates are opened, and the water flows through the turbines, turning electrical generators to make electricity. A 240-megawatt station has been operating in a small estuary in Brittany, France, for forty years, but it is the only example in the whole of Europe. [257]

In the UK, for example, if all exploitable estuaries were utilized in this way, some 15 percent of national electricity demand couldbe met, according to government estimates. [258]

To tap the energy of waves, a variety of devices have been designed, deploying turbines either on the shore, near-shore, or offshore in open water. Waves effectively concentrate the energy of wind: because water is much denser than air, the energy needed to move a certain volume of water is much greater than that needed to move the same volume of air.

For this reason, a small turbine past or through which a wave passes can generate the same power as a much bigger wind turbine.

A promising technology called Pelamis, under test offshore Scotland, consists of articulated cylindrical sections that are hinged so that they can move with the waves, powering as they do so hydraulic motors that generate electricity. A protorype 120 metres long and 3.5 metres wide generates 750 kilowatts.

A wave farm" of such devices spanning just a square kilometer of ocean would generate enough electricity for tens of thousands of homes. [259]
We are not space constrained when it comes to waves.
Similarly, currents in rivers can be tapped by small run-of-river turbines. Such technology is called micro-hydropower if the turbines are less than 1.25 megawatts.

There is huge potential here. It is quite amazing how much electricity can be generated by a line moving water.
In the UK for example, former mill sites alone have a combined generating capacity of one and maybe two typical nuclear or coal plants[260]

Biomass is an important renewable resource because of both its potential scale and the fact that it can be used in a way that produces no net greenhouse-gas emissions.
Biomass fuels are of three types: waste by-products (from agriculture, forestry and the urban environment), energy crops and processed fuels (for example wood pellets made from sawdust).
So long as the plant matter used is replaced by regrowth - hardly a problem when it comes to agriculture - there is no net build-up of greenhouse gases in the atmosphere.

Heat can be generated from these fuels by straightforward combustion in power plants and boilers, or by a group of higher-efficiency proesses: anaerobic digestion, gasification and pryrolysis.[261]

The heat from burning biomass can be used directly or used in a turbine to generate electricity. Again the potential is vast.

Take for example, production of straw in the UK.

Suppose one third of the annual production was burnt in biomass power plants: some 8 million tonnes of biomass material. This one crop residue would generate 3
percent of UK electricity.

Instead, most such residues simply go to waste.
As for energy crops such as fast-growing willow and Miscanthus there is strong potential.

In round figures, if 10 percent of the UK's 20 million hectares of agriculrural land were taken up with short rotation coppice (some 2 millionhectares) biomass could meet 10 percent of the UK's energy demand. Note that 0.6 million hectares of UK agricultural land is set aside as things stand ie essentially unused. [262]

Combined heat and power (CHP), as the name implies' allows any combustible fuel to be used with high efficiency.
Generating electricity alone in power plants is typically only 30-40 percent efficient because so much heat is lost.

CHP generators which use the heat produced as well as the electricity work at around 80 percent efficiency generating around three times more heat energy than electricity [263]
All power plants could also be CHP plants.

We don't have to be purist though. Biomass can, and is, burnt along with fossil fuel use and greenhouse-gas emissions.

This completes the renewable toolkit for power plants, large and small, in or distant from the buildings they are heating or lighting.

The renewable technologies that can be used within buildings are

68

often referred to by the catch-all term "micropower".

The renewable micropower family considers of both types of solar, micro-wind (small wind turbines on roofs), biomass boilers, micro-CHP (if the fuel is biomass) and ground-sourced heat pumps.

The role they play in electricity generation can be thought of as "embedded generation", in that they are embedded in the national electricity grid when the renewable micropower technologies are in common use' that grid will be a very different tanimal from the one that exists now. For the most part electrons flow in the grid one way only today: fiom the giant and dirty power plants wehave tended ro favour historically to rhe end user in their home, office, factory or whatever.

This requires high voltages, transmission lines running long distances, and transformer statrons to step the voltage down for use in buildings at the end of the line.

With embedded generadon, the distances travelled by the electrons is shorter, and everything is manageable with lower voltages.

Many electrons are used right there in the building where they are generated. Smaller grids can be used. The whole infrastructure tends to resemble the internet, where many "distributed"computers are used in preference to the giant centralized mainframes of old. More and more people are starting to refer to the "energy internet.[264]

It is all very well heating and electrifing buildings, but how are we going to get around in this brave new world? Once again, we are nor stumped for options, only the imagination to see them and the will to make the changes.

Automotive fuel for direct use can be made in abundance fiom plant matter.

Biodiesel can be made from such plant matter as soyabean, vegetable or rapeseed oil and used directly in cars with diesel engines.

In 2004, US manufacturers like DaimlerChrysler and General Motors suddenly began to take a serious inrerest in biodiesel. Volkswagen has said it will use biodiesel as the best way to compete with Toyota's immensely popular fuel-efficient hybrid car, the Prius.

Daimler Chrysler has said it will fill all new Jeep Liberry vehicles with biodiesel. [265]

Such inrerest can be explained at least in part by targets set by governments. Europe, for example, has a target for biofuels made from all agricultural forestry and organic waste: 2 percent of fuel use by 2005 and 10 percent by 2010.

Beyond biodiesel, there is ethanol – made from maize, and being produced in the US today in seventy-five small subsidised refineries (with a further twelve under construction).

On top of this, methane gas can be made, in principle in vast quantities, from anaerobic digestion of any organic waste, plant or animal, compressed and used as a fuel.

Plant matter can also be used rather rhan oil to make plastics and other chemicals important in energy technology.

At the 2003 Detroit motor show Ford introduced a concept called the Model U green car with engine oil made from sunflower seeds and seats made from soyabeans.

The company was merely extending a tradition: its founder Henry Ford, had used soyabeans rather than oil to make plastic and produced a car completely made from plants, as long ago as

1941. Such cars may never go on sale as things stand, but they could given the right conditions.

Their engines could be filled with fuel made fiom plants.

Biofuel refineries could be built faster, targets could be made more ambitious. Incentives of many kinds could be deployed to turn the Fords of this world from tinkerers with biofuelled biovehicles to mass producers. [266]
Then there is hydrogen, in which the auto giants are showing more than a passing interest.

Storage: the road to the hydrogen economy

Hydrogen is not a fuel in the strict sense, but rather an energy-storage medium, in that it is not found in narure unless combined with other elements. It is usable in fuel cells or as a solid or liquid fuel.

A fuel cell is a modular device that chemically recombines hydrogen with oxygen on a catalytic membrane producing an electric current as it does so plus a single and uniquely unproblematic waste product: pure hot water.

There are several different types of fuel cell, but they all work in this basic way[267]

The oxygen can typically come from the air, and the hydrogen can be made from any other energy source. At the undesirable end of the spectrum, because a lot of carbon dioxide is emitted along the way coal can be reformed (A simple way to think of reforming is taking the Hydrogen out of the hydrogen-and-carbon mix in hydrocarbons.). [268]

At the desirable end of the spectrum because no greenhouse-gas emissions result, renewable energy can be used for electrolysis to generate the hydrogen. (A simple way to think of this process is taking the H_2 out of the H_2O.) [268]

Fuel cells use their hydrogen about twice as efficiently as internal combustion engines use their gasoline, per unit of fuel. Currently however, they cost one hundred times more per unit of power and hydrogen itself is about five times as expensive as gasoline. [270]

General Motors, DaimierChrysler and Shell have all already invested more than a billion dollars each in fuel-cell research and development (R&D), trying to close this gap, and many other corporations have significant R&D programmes under way.
Governments are encouraging their national industries to view the search for commercial hydrogen vehicles as a race.
General Motors, endeavouring to set the pace for the US government, which has made a $1.2 billion R&D commitment over five years, has said publicly that it expects to begin selling hydrogen-fuel-cell vehicles in 2010.

It may have to hurry because the Chinese government is aiming to be the world leader in hydrogen-fuel-cell-powered cars and has been supporting R&D at the level of $200 million per year in the past few years. Local industries have so far produced more than a thousand patent applications in the area of fuel-cell technology. [271]

China is already believed to be the number two producer in the 50 million tonnes of hydrogen produced per year globally, and Japan has a specific goal of 50,000 fuel-cell vehicles on the road by 2010.
One hundred and seventy-two prototype hydrogen cars and eighty-seven hydrogen filling stations have been created worldwide thus far. [271]

Fuel-cell cars and buses are to be found today on the roads of many cities around the world. but not yet at prices affordable by most people.

The key to making this happen - the central thrust of the international R&D race is reducing their size and weight. In buildings that is less of a problem, and bulkier fuel cells are already to be found in pioneering green buildings, generating both electricity and useful hot water. [271]

Others have their doubts about fuel cells being the best way to use hydrogen. BMW for example, has opted for use of the gas in liquid or solid form - enabled by the storage under pressure of solids called
hydrides - and hydrogen gas-filling srations have already been built on an experimental basis in Germany. BMW in fact, began a massive marketing push in April 2001 to reposition itself as the lead pioneer of the hydrogen age.

Dr Helmut Panke, Chairman of the Board of Management, said the "*The hydrogen age has begun*".

Events like this make clear an important distinction between the auto and oil industries: the former is not necessarily going to lose its core product when the wheel of change really begins to turn.
In that, it is much less threatened than the oil industry. The oil industry will still be in business in the second half of the oil age, of course. It will be providing much-needed hydrocarbons for the chemical industry for one thing. But
it will be a shadow of its former self and, unless the companies take the lead in the renewable revolution, they will no longer be among the largest corporations in the world.

The challenge will be getting the auto industry to the point of no return, where they commit willingly - or are forced to commit - to the mass retooling of their factories for means of propulsion other than the internal combustion engine.

The success of the Toyota Prius in the last few years is forcing a major re-evaluation of attitudes to disruptive technologies among the major manufacturers. The Prius uses a combination of battery at low speeds and an efficient internal combustion engine at high speeds to achieve sixty miles per gallon of gasoline, compared to the record-low average of the 200 million vehicles in the American fleet last year running 20.4 miles per gallon. The popularity of the vehicle caught Toyota, by far the world's most successful auto manufacturer, by surprise.

The first model, introduced in 2000, sold only 15,000. As of September 2008, Toyota was selling that many a month of the roomier, more powerful model, with 22,000 customers remaining on waiting lists despite three ramp-ups of production.

Ford launched its first hybrid SUV in September 2004, and in 2007 there were some twenty-two hybrid versions of popular models. As General Motors' Vice Chairman Bob Lutz told *Newweek*:
"We can't iust sit there as a major corporation and say, 'Trust us, you'l lget a fuel cell from us and in the meantime we're not doing anlthing.'
With more and more of our competitors playing the hvbrid card, thereis just no way we can ignore that." [275]

Promising stuff, and it begs the question of just how much energy efficiency can reduce the target for renewables in replacing fossil fuels, both in the transport sector and in the built environment.

Efficiency: reducing the demand mountain to a hill

Arguably the biggest expert in the world on this subject is Amory
Lovins, the Director of the Rocky Mountain Institute in
Colorado.
Lovins himself regularly told an anecdote to make the point of
how energy efficiency is the forgotten cash cow.
A little girl walks down the street and sees a hundred-dollar bill
lying on the sidewalk. She says to her wise old grandad:
"There's a hundred-dollar bill on the ground!", But he says: "No
sweetie, if that was a hundred-dollar bill, someone would have
picked it up by now."

In 2004, Lovins and his team published the latest in their
exhaustive studies of the efficiency field, this time part-financed
by the pentagon if you please, and very much rooted in the
imperatives of our time. He called it *Winning the Oil Endgame:
Innovation for Profit, Jobs and Security.* [276]

Amazingly, but unsurprisingly to those who have exposed
themselves to a few details about energy efficiency, it concludes
as follows: "*... it will cost less to displace all the oil the United
States now uses, than it will cost to buy that oil.*"

To replace oil use with cheaper akernatives in this way, the
US would have to invest $180 billion over the next decade, for
which the return would be $130 billion in annual savings by
2025. To win this jackpot, the investrnent would need to be made
according to four technology strategies pursued step by step:
first, using oil twice as efficiently as is the case today; second,
substituting biofuels; third, saving natural gas; and fourth,
introducing hydrogen.
As will be clear from what we have considered above, these are
all trends that are under way already.
Doubling the efficiency with which the US uses oil can be

achieved with a variety of techniques, especially ultralight-vehicle design.

Lovins and his colleagues argue. Good as the efficiency of the current hybrid-electric cars is relative to the gas-guzzling norm, advanced composite or lightweight steel materials can nearly double that at an extra cost recoupable
from fuel savings in about three years.

Beyond the uptake of such smart technology, creative business models and public policies can easily make up the rest of the oil savings. A good example is the introduction of feebates: fees for inefficient vehicles or buildings that are rerycled in a revenue-neutral way as rebates for efficient ones.
The vehicle improvements and other savings don't need to break any records. They needn't even be as fast as those acrually achieved in America after the 1979 oil shock. The investment required to achieve this goal would be $70 billion of the total $180 billion.

Beyond efficiency savings of one half of all projected oil use a further quarter can be saved by creating a major domestic US biofuels industry.

Lovins and his team figure that rural America can be much strengthened by an assault by biofuels on hydrocarbon markets. Farm income could be boosted by tens of billions of dollars a year and more than 750, 000 new jobs created. This would require some $40 billion of the $180billion total investment.
The "low-hanging fruit" of efficiency savings in natural-gas use save at least half the gas demand projected by the US government in 2025.

The saved gas can be in part substituted for oil in a further demand reduction, or it can be recombined to make hydrogen displacing almost all the rest of US oil. The rest can easily be replaced by renewables.

Lovins make no radical assumptions about explosive growth of renewables or hydrogen.

Indeed, he classifies the development of hydrogen as optional. If you merely want to cut US oil consumption to the point that no imports are needed, you do the efficiency and biofuel parts. If you do the hydrogen and renewables part as well, you can wean the country off oil totally.

Some of the implications are mind boggling, especially the $180 billion investment needed to hit the plan in the context of the current high oil prices, nor to mention the $2.4 trillion investment needed to meet projected oil demand over the next decade according to Goldman Sachs, or the record cash hoards of the oil companies. [277]

Exxon alone has amassed $35 billion at the time of writing (in the face of investor pressure to shell some of it out as dividends, or go for a mega-acquisition). [278]

"The United States' economy already pays that much," Lovins observes of the $180 billion, "with zero return, every time the oil price spikes up as it has done in 2004."

Not to mention what will be paid for oil after the global production topping point.

How quickly can we withdraw from oil, gas and coal?

We can think of the renewable technologies as solar technologies in the broad sense: most are driven either direcdy or indirectly by sunlight falling on Earth. The sun creates differential warming in the atmosphere that creates winds, and so drives wind power. Winds create waves at sea, which drive wave-power devices. The sun creates differential warming in the ocean, which creates currents, which can drive marine turbines. Sunlight is needed for photosynthesis, and hence is at the root of plants useable as
fuel. The only type of renewable energy unrelated to sunlight in some way is tidal power, which is made possible by the gravitational pull of the moon.

We can think of the whole process of exploiting and commercializing these technologies as "solarization". This technologies themselves are renewable in the sense that they hold the potential for generating power as long as light falls on the planet: effectively, endless power.
It is far better to use all members of the solar-technologies family than to concentrate on one or two, obviously. The two main reasons for this are security in diversity and the ability to meet variable energy loads with spare capacity by mixing the means of supply. For example, solar can meet much of the peak air-conditioning load on summer days while biomass can contribute best to winter heating; marine can provide baseload while wind chips in where it can; and so on.
One of Shell's most famous scientists saw all this, a long time ago.
"Our culture doesn't know how to deal with a levelling off or a decline
but it will have to," said M. King Hubbert, "We have to steer ourselves into a stable state with as little catastrophe as possible. We should be looking for other sources of energy. There's only

one big enough. It's free, and it's good for at least a billion years' That's the sun" [279]

The UK government published a report in 2001 that concluded as follows: "... it would be technologically and economically feasible to move to a low carbon-emissions path, and achieve a virtually zero carbon-energy system in the long term, if we used energ'y more efficiently and developed and used low-carbon technologies."[280]

Among the low-carbon technologies on offer the government report places heavy emphasis on renewable energy and hydrogen, rather than nuclear power. Of solar energy the report concludes:
"[It] alone could meetworld energy demand by using less than I percent of land currently used for agriculture."

Former prime Minister Tony Blair used these same words in the speech he gave launching the UK Energy White Paper.

Faced with all the evidence of potential in the wider renuable/ storage/efficiency family what can we now say about how qurckly the world could solarize if it really wanted to?
Would we be fast enough to head off the economic trauma that an early oil topping point entails?

2. The shortfall between current expectation of oil supply availibilty and actual availibilty will be such that neither renewables, nor liquids from gas and coal, no nuclear nor any combination thereof, will be able to plug the gap in time to head off economic trauma as a result of the oil topping point.

Lead times

If the oil topping point happens this decade, even accepting the optimistic thoughts above, we are in trouble. "There really aren't any good energy solutions for bridges, to buy some time from oil and gas to the alternatives," says Matthew Simmons. "The only alternative right now is to shrink our economies." [286]

As John McGaughey of *World Energy Review*: puts it, "... anything that might be done to mitigate an oil reserves problem, such as oil shale or coal liquefaction or the hydrogen economy, is going to take twenty years or more to come on line." [287]
Conservation would take a decade or more.

It is easy to see why such views seem reasonable. The realization that growing supplies of cheap oil are no longer available will dawn on the energy traders at some point rather soon, and reductions in global oil supply of only a few percent in the past have been enough to trigger panic.

The first thing governments, industry and populations will do is look for energy-saving economies. Car-pooling programmes and bans on Sunday driving, may buy some time.
But with the early toppers projecting a 2 percent depletion per year against widely expected oil demand inreases of 2 percent and more, alternative supplies will very soon become imperative. And here they won't find big enough markets to plug the gaps, as things stand, by the end of the decade.

Never mind economics for the moment, let us just think about the timing. The time it takes to build liquids-from-gas plants, liquids-from coal plants, biofuel plants, hydrogen fuel plants, and hydrogen fuel-cell factories (before assembly in autos) is measured in years rather than months.[288]

OK, renewable micropower plants can be installed in as little as an afternoon, as solar installers have demonstrated hundreds of times over on roofs around the UK. But there is a problem here too: demand for solar PV and the other types of micropower kit has to be met at the factory gate, and factories don't go up in an afternoon.

It might take as long as 2 years to build a giant solar PV manufacturing plant from scratch, even with the stops pulled out.

Nuclear Inadequacies

In recent years the nuclear industry and its supporters In government have sought to position their technology for a revival after years in thedoldrums.

Nuclear is not the silver bullet for three main reasons: timing, investment and track record.

Timing

The timing problem is even worse with nuclear than it is with liquids from gas, liquids from coal, biofuels, hydrogen and fuel cells. In the UK nuclear expert Gordon MacKerron professes that "there is no realistic chance, given current politics. that nuclear power could deliver new power before about 2020"[289]

He draws this conclusion because in a country where no new reactor building has been agreed for fifteen years, no government could feasibly start a maior programme of construction without a major period of public consultation. Siting, licensing and local public-inquiry processes would take until around 2013. Only then could construction begin.

Historically, the Japanese have rushed through reactor-build programmes in five years, but more usually it has taken ten or more from planning to first power generation. By that time you would only be replacing old nuclear plants, many of which are at or beyond their planned lifetimes even today. In terms of overall national energy supply, this would require a major reactor-building programme just to stand still, as it were.
Just think what renewable energy and energy-efficiency markets could be doing by 2020, given even a fraction of the governmental and institutional support nuclear has been given for the last half-cenrury.
Investment

Most of the world's energ'y markers are liberalizing, and in liberalizing markets decisions about what electriciry generating plant gets built and what doesn't are ultimately made by investors, not governments.

As things stand - simply and graphically stated - financial institutions without exception ignore nuclear power as a repository for their invesrment dollars because of nuclear economics. Many opponents of nuclear begin
and end their case with this argument. Financiers simply won't prove persuadable that nuclear should be financed, they say, and it easy to see why. First, the long planning and construction times for nuclear mean a wait of at least seven years to see returns on capital invested, whereas with combined-cycle gas turbines (CCGTS) and large-scale wind, planning and building can happen in the order of a couple of years.

Second, the costs of CCGTs are a known entity in the markerplace.

Nuclear costs, with so many open liabilities from unsolved waste problems and potential accidents stretching out into the payback period, manifestly are not.

Third, total generating costs for nuclear are very sensitive to miscalculations with performance guarantees, given that 70 percent of the total generating cost involves up-front capital as opposed to capital for fuel and running costs. The proposed next generation of reactors have no proven track record, and the nuclear industry has a long history of over-promising and under-delivering in such matters.

CCGTs, meanwhile, come with performance guarantees backed by much operational experience, most of it to investors' liking.

Fourth, the nuclear industry needs economies of scale to make the proposed investment work. In the UK, for example, BNFL (British Nuclear Fuels Limited) argues that it needs ten giant reactors (ten l-giqawatt stations) to bring their costs down to competitive levels: one or two simply won't do it. [291]

I can concede that nuclear power is a low greenhouse-emissions technology. Even if nuclear power is truly a low-carbon technology, there is no point in having it if you don't need it in the first place because far more attractive options are available. Neither is there any point if you poison the world when you finally get to build your nuclear nirvana. To make any difference in reducing greenhouse-gas emissions. so many reactors would have to be built that the industry would be left with thousands of tonnes of plutonium to somehow handle and process safely.

A functional multilateral world with effective international law?

Suppose we have some warning of the approaching oil topping point. Matthew Simmons has pointed to transparency over oil reserves as a key requirement.

"I think we should basically look at this like we looked at
nuclear warfare and say that would be so awful if it happened -
let's do something, put in a warning system," he urges. [296]
Suppose governments listen to such enffeaties. Even then it is difficult to imagine a smooth transition from oil addiction to alternatives of any kind.

A greenhouse gas depletion prorocol, something that the town of Rimini in Italy is planning to adopt, could assist.

A Rimini Protocol might, for instance, require businesses to cut imports to match the world depletion rate of around 2 percent. A 2.5 percent a year reduction in imports is atainable. The result would be impressive: modest prices that even poor countries could afford, minimal energy needs met, and profiteering avoided. Best of all, consumers everywhere could be educated and prepared to face the reality of a changing world.
Is there a chance of such selfless collective thinking breaking out in the field of international relations? Given the experience of the Climate Convention, with its decade-plus of status-quo defence by vested interests, I am forced to conclude that the prospect is at best a very slim one.

The most probable outcome

The most likely outcome is that the world will drift on in overall collective denial. This book will be published, to join the other efforts under way to wake people up. It will be praised by some, vilified by others, but ignored by most people. At some stage as the topping point approaches, there will be a gathering upsurge of speculation that maybe the early-topping-point whistleblowers are on to something. Then thetsunami will hit.

The tsunami might be one giant wave of unstoppable panic and rumbling markets, like the Great Crash of 1929, or more than one wave, none giant but cumulatively as bad or worse than a single giant.

In the latter scenario, the oil price gets to a point that triggers recession. In that recession, as economic activity shrinks the low-hanging fruit of energy savings get picked as soon as possible.

Demand goes down as a result, the oil price goes down, so the economy improves. But demand for oil then goes up with the improving economies, and supply once again is inadequate - or perceived to be so - so the price goes back up and the economy dives again. The tsunami might therefore be a cyclical phenomenon of closely spaced oil-price peaks and recessions until the point that there is no more low-hanging alternative fruit, and the longrunshortfall between energy demand and supply becomes clear.

Major world events completely unrelated, or only indirectly related, to the oil topping point will inevitably muddy this picrure. There are more reasons than the price of oil for global markets to crash, as notorious speculator George Soros often points out.

It is easy to think of possibiiities that might intersect with the playing out of the oil-depletion drama in ways that could accelerate or delay, amplify or suppress, a fusing of the psychological peak panic point with the physical oil-production topping point. An acceleration/amplification might be the fall of the House of Saud to a fundamenralist regime. A delay/suppression might be a major setback for the chinese economy. I am not going to speculate here about the detail of how the crisis will play out. or what the years immediately after the tsunami - whatever form it takes - will look and feel like sector by sector across society and economy.

Suffice to say that it doesn't look very pretty, however you look at it.
The main point is this. In the wake of the realization of the oil production topping point, the onus in the energy sector will be on immediate damage limitation. Energy services and renewable energy companies will be besieged. Politicians will want to know how quickly we can accelerate if they finally give us as much support as they have given nuclear and the military pursuit of oil-supply protection all these years.

Corporations will want to cut deals with us exclusively to keep their lights on and not their competitors. Consumers will suddenly be desperate to be taken off the national grid and given heat and electriciry at almost any price.

Come whatever in other societal and economic sectors, people working in renewable energy, energy storage and energy efficiencv will be in the front row of those who can help once widespread acceptance of the oil topping point and its implications has descended on the world.

3. Renewable energy and fuel use, alongside energy efficiency will increasingly substitute for oil and gas growing explosively whatever happens.

The first major financial institution to set up an investment fund for renewable energy suffered something of a disaster. Merrill Lynch's £200 million New Energy Fund floated shortly before the dot.com crash of 2000 and was one of the many victims trailed in its wake.

This was a painful experience for many people. Big money was put off for years. But in 2004, with the dot.com bear market over, oil prices rising, and the prospect of the Kyoto Protocol coming into force, a number of major financial institutions began setting up investment funds for renewable energy, and that trend has accelerared. Just as in the run-up to the dot.com boom, the business magazines finally began to take notice.

Business Week asked on its front cover: "`Global warming: why is business taking it so seriously?" [297]
Fortune went so far as to draw up its own plan for a renewables revolution. [298]

Veteran investors began to talk of the dangers of a boom in renewables investment, and even a dot.com-style feeding frenzy.

The solar PV industry serves as a good example. The journal *Photon* records an index of fourteen quoted companies whose businesses are more than 50
percent solar PV. In 2004, this index rose 182 percent. In contrast, and notwithstanding a persistently high oil price, the index for the twelve largest oil stocks rose only 18 percent. [300]

Of course, the fourteen solar companies have a collective market capitalization of not much more than $1 billion, compared to goodness knows how many hundreds of billions for the oil giants, but investors like growth. That is where they go first for investment options.

The Photon index excludes the solar companies with giant parents doing mostly other things for the moment, such as Sharp and Sanyo in electronics and BP and Shell in oil. It also excludes ali the private companies.

When all these are added in, the embryonic PV industry doesn't look so embryonic after all.
In July 2004, Credit Lyonnais Securities Asia conducted the most exhaustive sector study yet of the PV industry. They found a market of more than $7 billion growing fast at 30 percent per year, and an overall profit pool of $800 million.

They expect the industry's growth to reach $30 billion in revenue and more than $3 billion in profit by 2010. As they worked their way through more than two hunclred solar companies, the CLSA analysts started out sceptical
and ended up enthusiastic. In their writing, they can barely contain their surprise.

"Lost in the noise has been a deeper, fundamental story'," they conclude. "Solar power is hot." [301]

One reason for this is that solar power can compete with retail prices, not generator *costs*. This an important point to bear in mind when it comes to personal or collective action In the coming energv crisis.

Solar PV is a unique technology generating right where the power is needed primarily in buildings and so bears direct comparison with the retail price of power, including everything the utilities load into those prices: their generating costs in conventional polluting power stations, transmission and distribution costs in the grid, taxes, exorbitant profits and anything else they can get away with.

Yet time and again energy analysts compare the costs of a PV system with only the generating costs of coal, nuclear, gas, oil or wind.
No wonder people say solar is "too expensive" (three to ten times more so, on this unfair plaving field).
The CLSA analysts have one word for this: "irrelevant".
Another reason for the soaring investor interest in solar is that PV costs are coming down at 5 percent per year on average over the last few years, while the price of oil and gas continues to search for the roof.

Solar manufacturers have significant economies of scale to look forward to as they scale up. Also, they apply knowledge gleaned at low volumes to achieve efficiencies at higher volumes.

For this reason, costs decrease by around 20 percent for each doubling of capacity - a long-term trend that
is expected to continue in the years ahead.

In fact, there is potential for costs to come down even faster. Meanwhile, what is oil doine but generally going up? And where oil prices go, gas tends to follow.

A further reason is that solar power is alreadv more economical than the polluting alternatives in many markets. In Japan and Germany for instance, the two largest markets for PV government market-building incentives have made the price of solar power competitive with the residential grid power price by, respectively, applying direct subsidies
and using a premium buyback rate.

The CLSA analysts conclude:

"Our initial reaction to solar's dependence on incentives was to discount the potential of solar power. Our view has changed as we became more convinced that incentives result from perceived global climate change risks and energy security/price concerns that are unlikel,v to disappear anytime soon." Moreover, the Japanese subsidies are not being phased out, and yet there is no sign of the market slowing'

This sort of interest in the rmoney world for renewables is real and growing. It is causing the renewables industries to explode up hockey-stick growth curves. They can become very big, very quickly. The longer it takes for the full weight of the energv crisis to hit, the better they will be able to limit the damage.

But this burst of optimism has to be severely tempered. The oil and

gas industrv is vastly bigeer. So to is the coal industry.

4. Amid the ruins of the oil energy modus operandi, many will try to turn to coal, and so the extent to which renewable energy grows explosively instead of coal expansion, rather than along side it, will determine whether economies and ecosystems can survive the climate change threat.

The coal industry is strangely hardline. It utilizes a technology that is so clearly damaging and yet it continues to grow largely unapolgetically around the world.

Soon will come the clarion call, no doubt in many countries, for gasoline to be made by coal gasification.

People will be looking anwyhere they can for gasoline substitutes once the oil topping point is upon us. Squeezing oil out of coal is one method for doing that costly and environmentally disastrous though it may be.

The prospect of widely attainable sequestration technoloqies will deepen the temptations to go all-out for coal.

If there is even a slight mismatch between the amount of carbon burnt in coal and the amount of carbon in carbondioxide emitted instead of sequestered - as there surely would be given the scale of the current demand and the embryonic state and riskiness of the sequestrarion technologies - we would soon burst through the danger threshold of a 2°C increase in the global average temperature and soar beyond.

Sequestration of carbon dioxide from coal

Let us take a quick tour ofcurrent research and development in sequestration to give a feel for the scale of the challenge facing advocates of renewable and efficient-energy technologies when it comes to the matter of coal. Proponents envisage three types of sequestration: geological (the injection of carbon dioxide into rock strata below ground) oceanic (the pumping of carbon dioxide into the oceans) and biologicai (the simulated soaking up ofcarbon dioxide by plants onland or in the sea).

When it comes to geological sequesffation' the United States is leading the way. As US Energy Secretary Spencer Abraham put it in 2003, "... *carbon sequestration has rapidly grown in importance to become one of the Administration's highest priorities. Our activities and our plans bear out the determination with which we arc pursuing the promise.*"

Accordingly, the US Department of Energy's sequestration programme has several active research projects under way In Canada New Mexico, Virginia and Texas, carbon dioxide is already being injected, or will be injected if the plans go live, into one or more of the three main underground storage options: depleted oilfields, unmineable coal seams and deep saline

By 2009, the DoE's aim is to initiate at least one large-scale demonstration of carbon-dioxide storage (>1 million tonnes per year) in a geological formation.

Estimates of sequestration costs based on the technologies available today are in the range of $100 to $300 per tonne of carbon emissions avoided. The goal of the sequestration programme is to reduce the cost of carbon sequestration to $10 or less per net tonne of carbon emissions avoided by
2015. By 2050, the Department professes that this technique will be saving more carbon-dioxide emissions than renewables and efficienry combined. [306]

The European Union is collaborating enthusiastically with this research, and has significant numbers of programmes of its own in the works. The UK government has produced a report saying that largescale sequestration may be needed to help it reach a target of 20 percent greenhouse-gas emissions' reductions by 2020. However, in Europe at least some scrutiny of the potential downsides of sequestration is likely.

The same report conciuded "... it is currentiy impossible to analyse with any confidence the likelihood ofaccidental releases from carbon dioxide sequestration reservoirs". [307]

It further raises questions about the legal liability of companies if gases do escape, and the insurability of
operations on a large scale.

Proponents of oceanic sequestratron argue that carbon dioxide from power plants could be piped across land, and the continental shelves, and then pumped into the deep oceans. The argurnent is that the gas will dissolve in sea water at depth, and not be returned to the surface for hundreds of years. [308]

Such an approach would contravene the internationaltreaty controlling waste disposal at sea, the London Dumping Convention.

Moreover, major concerns about acidification of the oceans ought to render this idea a non-starter. It appears from the most detailed measurements yet made, published in 2007, that the oceans have absorbed almost half oi all carbon dioxide from fossil-fuel burning and cement manufacturing to date. "The oceans are producing this tremendous service to humankind by reducing the amount of carbon dioxide in the atmosphere," savs Chris Sabine, an oceanographer at the US National Oceanic and Atmospheric Administration in Seattle who has been a leader in this research. "But it's changing the chemistry of the oceans."

The research results are based on ten years of criss-crossing of the globe by ships making nearly 10,000 stops for measurements along the way. The core problem is that carbon dioxide dissolves in sea water to form carbonic acid, which in turn can dissolve the shells and skeletons of marine life. [309] Aside from general ecological concerns given that fish provide humans with most of their protein intake this is clearly problematic.
Beyond geological and oceanic sequestration, others have proposed an all-out effort to boost biological sequestration.

We have known that carbon dioxide can be taken out of the atmosphere and into fast-growing plants for as long as photosynthesis has been understood.
A hectare of immature forest can absorb more than 100 tonnes of carbon each vear. [310]

But recent detailed studies have shown that the effect is not as significant as biologists once thought. Certainly it is nowhere near big enough to offset emissions from fossil-fuel burning. [311]

Undeterred, some scientists have proposed boosting the amount of carbon dioxide taken out of the
atmosphere by increasing the numbers of phytoplankton, single-cell plants that float by the mryriad in surface waters of the world's oceans.

This could be done, they say, by doping the oceans with iron, so boosting nutrient for the phytoplankton, causing them to bloom [312].

This is yet another example of a proposed reckless gamble in a poorly understood and visibly stressed environment, when perfcectly safe alternatives are available in volume in renewable and efficient energy technologies.

The bottom line is this. Two ideas will confront each other head-on in the wake of the oil topping point. We can call them solarization and coalification. This I contend will be the battleground that will decide the fate of the planet.

Hope lives and the race is on.

NOTES

Notes
Pages 1 to 6

1. Chris Skrebowski, "Joining the dots" presentation to Energy institute conference "Oil Depletion: no problem, concern or crisis?" London 10th November 2004.

2. A barrel of oil contains 42 US gallons, weighs 0.1364 tonnes and can fill a typical gasoline tank several times. The six barrels comes from "The Price of Steak" National Geographic, June 2004, p. 98. The article cites a 1250 lb steer requiring 283 gallon. I barrel= 42 US gallons, 6 barrels= 252 gallons@30 miles per gallon= 7,590 miles. Distance between New York and LA is 2,800 miles.

3. US Energy Information Administration

4. In the IEA's World Energy Outlook 2004 reference case, world oil demand increases by 1.6 percent annually to 121 million barrels per day in 2030.

5. www.eia.doe.gov/emeu/cabs/pgulf.html

6. www.fueleconomy.gov/feg/FEG2005GasolineVehicles.pdf

7. Michael Klare, Blood and Oil, 2004, p46.

NOTES
Pages 9 to 36

1 Deffeyes KS (2006), Join us as we watch the crisis unfolding, 11 February 2006 [accessed 17 June 2008]

2
 Deffeyes KS (2005), Beyond oil: The view from Hubbert's peak, p. xiii, Princeton: Princeton University Press

3
 Campbell CJ (2002), Forecasting global oil supply 2000-2050, Hubbert Center Newsletter No. 2002/3

4
 Anon (2006), Total sees 2020 oil output peak, urges less demand, Reuters, 7 June 2006.

5
 Hubbert MK (1949), Energy from fossil fuels, Science, 109 (2823), 4 February 1949, pp. 103-109

6
 Hubbert MK (1956), Nuclear energy and the fossil fuels. API Conference, San Antonio, TX (March 7–9, 1956), later published as
Publication no. 95, Shell Development Company (June 1956), p. 26

7
 Deffeyes K (2001), Hubbert's peak: The impending world oil shortage, Princeton: Princeton University Press

8
 Strahan D (2007), The last oil shock, London: John Murray, p. 44

9
 Laherrere J (2002), Comments on the book: Hubbert's peak: The impending world oil shortage, 6 January 2002 [accessed 17 June 2008]

NOTES
Pages 9 to 36

10
Hubbert MK (1980), Techniques of prediction as applied to the production of oil and gas; Oil & Gas Supply Modeling, Ed. S.I. Gass;
Proceedings of a symposium held at the U.S. Department of Commerce, National Bureau of Standards, Washington, D.C., June 18-20,
1980; Report N.B.S. Special Publication #631, May, 1982, p. 16-141
11
Hubbert Peak of Oil Production (undated), Theory [accessed 17 June 2008]
12
Campbell CJ (2002), Forecasting global oil supply 2000-2050, Hubbert Center Newsletter No. 2002/3
13
Conventional oil excludes oil from coal and shale, heavy oil, deep-water oil (>500m), and polar oil
14
Bartlett AA (2000), An analysis of US and world oil production patterns using Hubbert-style curves, Mathematical Geology, 32 (1), pp. 1-17
15
Campbell CJ (2002), Forecasting global oil supply 2000-2050, Hubbert Center Newsletter No. 2002/3
16
Deffeyes KS (2006), Join us as we watch the crisis unfolding: Estimate of uncertainty, 14 June 2006 [accessed 17 June 2008]
17
Robelius F (2007), Giant oil fields – the highway to oil. Giant

NOTES
Pages 9 to 36

oil fields and their importance for future oil production. Uppsala dissertations from the Faculty of Science and Technology 69. 156pp
18
Skrebowski C (2006). Quoted in Strahan D (2007), The last oil shock, London: John Murray, p. 203.
19
US Geological Survey (2000), World energy assessment [accessed 17 June 2008]
20
Energy Information Administration (2004), Long-term world oil supply scenarios: The future is neither as bleak nor as rosy as some
assert, 18 August 2004 [accessed 17 June 2008]

21 ibid

22 National Petroleum Council (2007), Hard truths: Facing the hard truths about energy, July 2007, p. 115

23 Energy Information Administration, Table 1.1d World crude oil production (including lease condensate), 1997-present, 9 June 2008
[accessed 17 June 2008]

24 Deffeyes KS (2008), Join us as we watch the crisis unfolding: The second Great Depression, 6 February 2008 [accessed 17 June
2008]

98

NOTES
Pages 9 to 36

25. Energy Information Administration, Table 1.1d World crude
oil
production (including lease condensate), 1997-present, 9 June
2008
[accessed 17 June 2008]

26 Energy Information Administration, Table 4.1d World crude
oil production (including lease condensate), 1970-2007, 9 June
2008
[accessed 17 June 2008]

27 National Petroleum Council (2007), Hard truths: Facing the
hard truths about energy, July 2007, p. 18
"Minerals are inexhaustible and will never be depleted. A stream
of investment creates additions to
proved reserves, a very large in-ground inventory, constantly
renewed as it is extracted... How much
was in the ground at the start and how much will be left at the
end are unknown and irrelevant"

28 .Adelman MA (1993), The Economics of Petroleum Supply.
Boston: Massachusetts Institute of Technology, p. xi
4

29 . Goodstein D, quoted in Semple RB (2006), NY Times: The
end of oil, 28 February 2006, [accessed 11 June 2008]
30
US Geological Survey (2000), World energy assessment

NOTES
Pages 9 to 36

31 .For example, Campbell CJ (2002), Forecasting global oil supply 2000-2050, Hubbert Center Newsletter No. 2002/3
Deffeyes KS (2005), What happens once the oil runs out?, New York Times, 25 March 2005

32 .US Geological Survey (2000), World energy assessment

33. Campbell CJ (2002), Forecasting global oil supply 2000-2050, Hubbert Center Newsletter No. 2002/3

34 ibid
35. International Energy Agency (undated), About the IEA [accessed 17 June 2008]
36. International Energy Agency (2007), IEA response system for oil supply emergencies, 2007, p. 15
37. ibid, p. 3
38. ibid, p. 7
39. ibid, p. 11
40. ibid

41.International Energy Agency (2005), Resources to reserves: Oil and gas technologies for the energy markets of the future. Paris:
International Energy Agency, p. 13
42
 International Energy Agency (2007), World Energy Outlook 2007: Fact Sheet – Oil
43
 National Petroleum Council (2007), Hard truths: Facing the hard truths about energy, July 2007, p. 5

NOTES
Pages 9 to 36

44 Semple RB (2006), NY Times: The end of oil, 28 February
2006, [accessed 11 June 2008]
45
ExxonMobil (2006), Peak oil?   Contrary to the theory, oil
production shows no sign of a peak [accessed 18 June 2008]
6
46
McNulty S, Exxon oil production struggles for growth, Financial
Times, 2 May 2008
47
National Petroleum Council (2007), Hard truths: Facing the hard
truths about energy, July 2007, p. 91
48
BBC (2004), Oil giant Shell's investors shocked, 15 July 2004
[accessed 17 June 2008]
49
The SEC permits oil companies to disclose only proved reserves
that a company has demonstrated by actual production or
conclusive
formation tests to be economically and legally producable under
existing economic and operating conditions
50. Shell (2004), Proved reserves recategorisation following
internal review: No material effect on financial statements, 9
January 2004
[accessed 12 May 2008]
51 .BBC (2004), Oil giant Shell's investors shocked, 15 July
2004 [accessed 17 June 2008]
52 Webb T (2008), Shell to write off half of last year's reserves,

NOTES
Pages 9 to 36

53. van der Veer J, quoted in Mortishead C (2008), Shell chief fears oil shortage in seven years, The Times, 25 January 2008

54. Hofmeister J, quoted in Hargreaves S (2008), Don't blame us for prices – oil execs, CNNMoney.com, 20 May 2008

55. Malone R, quoted in Hargreaves S (2008), Don't blame us for prices – oil execs, CNNMoney.com, 20 May 2008

56 .Yergin D (2005), It's not the end of the oil age, Washington Post, 31 July 2005

57
Morton G (2008), Holding Daniel Yergin and CERA accountable, 10 January 2008 [accessed 17 June 2008]

58
Riva JP (1999). Is the world's oil barrel half full or half empty? It depends upon whether you're an economist or a geologist!, Hubbert
Center Newsletter No. 99/2

59
Udall R and Andrews S (1999). When will the joy ride end? A petroleum primer, Hubbert Center Newsletter No. 99/1

60
Deffeyes KS (2005), Join us as we watch the crisis unfolding, 28 November 2005,

61
Tweed D (2008), Greenspan says oil to keep rising on capacity limits, Bloomberg, 14 May 2008 [accessed 17 June 2008]

62
Macalister T (2008), Brown wants profits poured into North Sea, The Guardian, 30 April 2008

63
Bodman S (2008), Testimony to the House Select Committee on

NOTES
Pages 9 to 36

Energy Independence and Global Warming, US Department of
Energy, 22 May 2008

64

European Commission press release IP/08/916 (2008),
Commission calls for swift adoption of energy and climate
policies as best
coordinated response to rising oil prices, 11 June 2008

8

65

Simmons MR (2002), The world's giant oilfields, 9 January
2002 [accessed 17 June 2008]

66

ibid

67

ibid

68

Energy Information Administration (2007), International energy
outlook 2007, May 2007

69

ibid, p. 12

70

ibid, p. 30

71

ibid, pp. 14 and 30

72

Actual world oil price is 2008 average, up to and including 14
July 2008

73

Petroleum refers to all conventional crude oil and energy liquid
substitutes

74

Energy Information Administration (2007), International energy

NOTES
Pages 9 to 36

outlook 2007, May 2007, p. 29

75
Proved reserves are estimated quantities that analysis of geologic and engineering data demonstrates with reasonable certainty (80-90 per cent) are recoverable under existing economic and operating conditions.
76
Energy Information Administration (2007), World proved reserves of oil and natural gas, most recent estimates, 9 January 2007
[accessed 17 June 2008]
77
For example, Robelius F (2007), Giant oil fields – the highway to oil. Giant oil fields and their importance for future oil production.
Uppsala dissertations from the Faculty of Science and Technology 69. 156pp,
Riva JP (1999). Is the world's oil barrel half full or half empty? It depends upon whether you're an economist or a geologist!, Hubbert
Center Newsletter No. 99/2, and
Ivanhoe LF (2000), Oil reserve revisions: Major OPEC and communist countries 1979 to 1999, Hubbert Center Newsletter No. 2000/2-2,
and
Laherrere J (2007), Uncertainty of data and forecasts for fossil fuels [accessed 17 June 2008]
78
Riva JP (1999). Is the world's oil barrel half full or half empty? It depends upon whether you're an economist or a geologist!,

NOTES
Pages 9 to 36

Hubbert
Center Newsletter No. 99/2
79
Anon (2006). Oil reserves accounting: The case of Kuwait,
Petroleum Intelligence Weekly, 30 January 2006
80
Ivanhoe LF (2000), Oil reserve revisions: Major OPEC and
communist countries 1979 to 1999, Hubbert Center Newsletter
No. 2000/2-
2, p. 6
81
ibid
82
Saudi Arabia, Kuwait, Iraq, Iran, United Arab Emirates, Russia
and China
83
Ivanhoe LF (2000), Oil reserve revisions: Major OPEC and
communist countries 1979 to 1999, Hubbert Center Newsletter
No. 2000/2-
2
84
ibid
85
Wicks M (2006). Peak oil letter from UK Energy Minister,
Energy Bulletin, 9 May 2006
86
10 Downing Street (2007), Peakoil – epetition reply [accessed
17 June 2008]
87
Brown G (2008), Gordon Brown: We must all act together, The
Guardian, 28 May 2008

NOTES
Pages 9 to 36

88
DTI (2003), Energy White Paper: Our energy future – creating a low carbon economy, London: DTI, p. 9
89
Tran M (2005), Q&A: Gas Prices, The Guardian, 23 November 2005
90
DBERR (2008), Table 3.1.1: Crude oil and petroleum products: Production, imports and exports 1970 to 2006 [accessed 26 June 2008]
11
91
Hubbert MK (undated), Handwritten note on Ivanhoe's copy of NBS special publication 631, 1982, pp. 140-141
92
BP (2007), Oil: Spot crude prices [accessed 18 June 2008]
93
Energy Information Administration (2008), Crude1 [accessed 18 June 2008]
94
Robelius F (2007), Giant oil fields – the highway to oil. Giant oil fields and their importance for future oil production. Uppsala dissertations from the Faculty of Science and Technology 69. 156pp
95
ibid
12
96
Saudi Arabia Market Information Resource and Directory (2004), Saudi Arabia to raise oil production to 11 million barrels per day, 28
September 2004 [accessed 17 June 2008]

NOTES
Pages 9 to 36

97
Royal Embassy of Saudi Arabia (2005), Oil Minister address to the 18th World Petroleum Congress, 10 March 2005 [accessed 17
June 2008]
98
Energy Information Administration, Table 1.2: OPEC crude oil production (excluding lease condensate), 1997-present, 10 July 2008
[accessed 16 July 2008]
99
al-Huseini S (2007), Former head of Saudi Aramco: Oil has peaked [accessed 17 June 2008]
100
Birol F, quoted in Mouawad J (2008), Oil price rise fails to open tap, New York Times, 29 April 2008
101
Staff correspondents (2008), Oil prices dive, The Australian, 3 May 2008
102
McRae H (2008), We will never have cheap oil again, The Independent, 30 April 2008,
103
Rubin J and Buchanan P (2008), How much higher will oil prices go?, CIBC World Markets, 24 April 2008,
104
Harding J (2008), Crude forecast to reach $225US, The Calgary Herald, 25 April 2008
105
Goldman Sachs (2008), Global: Energy: Oil [accessed 17 June 2008]

NOTES
Pages 9 to 36

106
BP (2007), Oil: Spot crude prices [accessed 18 June 2008]
107
Energy Information Administration (2008), Crude1 [accessed 18 June 2008]
108
Government Printing Office (2006), Gross Domestic Product and deflators used in the historical tables: 1940-2009 [accessed 18 June 2008]
109
Up to and including 18 June 2008
110
Deffeyes KS (2008), Join us as we watch the crisis unfolding: The New York Times, 6 March 2008 [accessed 17 June 2008]
14
111
Oil consumption includes inland demand plus international aviation, marine bunkers, and oil products consumed in the refining process
112
Energy Information Administration (2007), World proved reserves of oil and natural gas, most recent estimates, 9 January 2007 [accessed 17 June 2008]
113
BP (undated), World oil consumption [accessed 17 June 2008]
15
114
IEA (2008), Oil market report, 10 June 2008

NOTES
Pages 9 to 36

115
Sandrea I and Barkindo M (2007), West Africa-1: Undiscovered oil potential still large off West Africa. Oil & Gas Journal 105 (2), 8
January 2007
116
Radler M (2006), Special report: Oil production, reserves increase slightly in 2006. Oil & Gas Journal 104 (47), 18 December 2006
117
Robelius F (2007), Giant oil fields – the highway to oil. Giant oil fields and their importance for future oil production. Uppsala dissertations from the Faculty of Science and Technology 69. 156pp
118
ibid
119
Information from BP, Production [accessed 18 June 2008]

121
Hoyos C and Blas J (2008), Fears emerge over Russia's oil output, Financial Times, 14 April 2008,
121
Energy Information Administration (2008), Table 4.1c World crude oil production (including lease condensate), 1970-2007, 9 June
2008
16
122
Energy Information Administration (2007), Table 4.1c: World crude oil production (including lease condensate), 1970-2007, 9

NOTES
Pages 9 to 36

June
2008 [accessed 1 July 2008]
123
Ivanhoe LF (2000), Petroleum positions of the United Kingdom
and Norway Western Europe, Hubbert Center Newsletter No.
2000/2-
1
124
ibid
125
DTI (2007), Meeting the energy challenge: A White Paper on
energy. London: DTI, p. 4
126
Ivanhoe LF (2002), Canada's future oil production: Projected
2000-2020, Hubbert Center Newsletter No. 2002/2
127
Rubin J and Buchanan P (2008), How much higher will oil
prices go?, CIBC World Markets, 24 April 2008
17

128
BBC, Indonesia to withdraw from OPEC, 28 May 2008
[accessed 18 June 2008]
129
Members' Research Service calculation, based on Annex A
130
ibid
131
Rubin J and Buchanan P (2008), How much higher will oil
prices go?, CIBC World Markets, 24 April 2008

110

NOTES
Pages 9 to 36

132 .ibid
133 .ibid
134
Rubin J (2008), The age of scarcity, CIBC World Markets Inc.
135
IEA (2008), Oil market report, 13 May 2008
136
Anon (2008), IEA trims world oil demand, cuts supply forecast, Reuters, 10 June 2008
137
IEA (2008), Oil market report, 13 May 2008, p. 3
138
ibid, p. 16
18
139
ibid, p. 14
140
ibid, p. 15
141
Burton J, Malaysia petrol prices rise 40%, Financial Times, 5 June 2008
142
See, for example, The Associated Press, Venezuela to sell oil to Portugal in exchange for food, technology, Chavez says, International Herald Tribune, 14 May 2008
143
US-China Economic and Security Review Commission, 2006 report to Congress, November 2006, p. 102
144
ibid, p. 107

NOTES
Pages 9 to 36

145 .ibid

146

IEA (2008), Oil market report, 13 May 2008

147

EU Business (2008), European Parliament calls for biofuel targets to be cut, 8 July 2008 [accessed 9 July 2008]

148

Traynor I (2008), EU set to scrap biofuels target amid fears of food crisis, The Guardian, 19 April 2008

149

The Royal Society (2008), Sustainable biofuels: Prospects and challenges, Policy Document 01/08, London: The Royal Society

150

Environmental Audit Committee (2008), First report: Are biofuels sustainable?, 15 January 2008

151

International Transport Forum Press Release (2008), Secretary General of the International Transport Forum calls for low-carbon

transport system, 30 May 2008

152

UK Petroleum Industry Association (2008), Briefing: Renewable Transport Fuels Obligation (RTFO), April 2008

153

Data sets from 2002 to 2004 inclusive are regarded as 'experimental'; subsequent datasets have been approved by National

Statistics

19

May 2006 [accessed 17 June 2008]

112

NOTES
Pages 9 to 36
159
Hirsch RL et al. (2005), Peaking of world oil production:
Impacts, mitigation and risk management, February 2005, p. 7
[accessed 17
June 2008]
160
Regeringskansliet [Swedish Government], 2005.   Strategic
challenges: A further elaboration of the Swedish strategy for
sustainable
development [accessed 17 June 2008]
21
161
Hubbert Peak of Oil Production (undated), Swenson's law
[accessed 17 June 2008]
162
Deffeyes KS (2005), Beyond oil, p. 181. New York: Hill and
Wang, 202pp
163
Energy Alternatives (undated), Swenson's law [accessed 17 June
2008]
164
International Transport Forum Press Release (2008), Secretary
General of the International Transport Forum calls for low-
carbon
transport system, 30 May 2008
165
Campbell CJ (2002), Forecasting global oil supply 2000-2050,
Hubbert Center Newsletter No. 2002/3
166
Schlesinger J (2005), Statement of James Schlesinger before the
Committee on Foreign Relations, United States Senate, 16
November 2005 [accessed 17 June 2008]

NOTES
Pages 9 to 36

167
Strahan D (2007), The last oil shock, London: John Murray, p. 123.
168
Oil Depletion Analysis Centre (undated), Preparing for peak oil, p. 2
169
ibid
170
Fortson D (2008), An ominous warning that the rapid rise in oil prices has only just begun, The Independent, 11 June 2008
171
Warner J (2008), The oil price will eventually return to earth, but collateral damage is likely to be serious, The Independent, 23 May
2008
172
Oswald A (2000), Oil and the real economy: An interview with Andrew Oswald, Warwick University interview, 17 March 2000 [accessed 17 June 2008]
22
173
ibid
174
Hirsch RL et al. (2005), Peaking of world oil production: Impacts, mitigation and risk management, February 2005, p. 7 [accessed 17
June 2008]
175
Hirsch RL et al. (2005), Peaking of world oil production: Impacts, mitigation and risk management, February 2005, p. 7 [accessed 17 June 2008]

NOTES
Pages 9 to 36
176
ibid, p. 6
177
Hirsch RL et al. (2005), Peaking of world oil production:
Impacts, mitigation and risk management, February 2005, p. 7
[accessed 17
June 2008]
178
Conway E (2008), Growth forecast for UK economy will be hit
by rising oil price, The Daily Telegraph, 10 May 2008
179
Tanaka N, quoted in TradeArabia (2008), Oil price can trigger
recession: IEA, 22 April 2008 [accessed 17 June 2008]
180
Westcott RF (2006), What would $120 oil mean for the global
economy?, Securing America's Future Energy, April 2006
23
181
Campbell CJ (2002), Forecasting global oil supply 2000-2050,
Hubbert Center Newsletter No. 2002/3
182
ibid
183
Rees-Mogg W (2008), King oil will turf out Gordon Brown, The
Times, 12 May 2008
184
Kaletsky A (2008), Could oil mania be coming to an end?, The
Times, 1 May 2008
185
Conway E (2008), Oil's surge to $120 poses new threat to UK
economy, The Daily Telegraph, 10 May 2008

NOTES
Pages 9 to 36

186
Energy Information Administration (2007), International energy
outlook 2007, May 2007, p. 13
187
 ibid, pp. 14-15
188

NOTES
Pages 37-40

1
Meadows D. et al 1972: *"The Limits to Growth, The Club of Rome"*, by Donella H. Meadows, Dennis l. Meadows, Jorgen Randers, William W. Behrens III.

2
Simon J. 1980: "Resources, Population, Environment: An Oversupply of False Bad News," Science, 208, June 27, 1980, pp. 1431-1437pp. 1431-1437.

3
Paul Ehrlich, 1968: "The Population Bomb" Random House, (January, 2000 edition).

4
Simon Erlich bet: See for example: the entry in the Wikipedia, http://en.wikipedia.org/wiki/Julian_Simon

5
BP Statistics 2008: *"Statistical Review of World Energy 2005"*, http://www.bp.com Published on 14th June 2008

Notes
Pages 41 to 59

7.
L.F. Ivanhoe, "Updated Hubbert Curves analyze world oil supply," World Oil, November
1996, pp. 91-94. C.J. Campbell and J.H. Laherrere, "The End of Cheap Oil," Scientific
American, March 1998, p. 78-83. J.H. Laherrere, "World oil supply—what goes up must
come down, but when will it peak?" Oil & Gas Journal, February 1, 1999, p. 57-64. .

8.
 Archived at
http://www2.exxonmobil.com/corporate/newsroom/publications/thelampno1_2003/page_5.
html

9
. Michael Ruppert, "Colin Campbell on Oil," From the Wilderness (October 23, 2002).
Archived at
http://www.fromthewilderness.com/free/ww3/102302_campbell.html

Notes
Pages 41 to 59

10 . David Pimentel and Mario Giampetro, "Food, Land, Population and the US Economy" Carrying Capacity Network, 11/21/1994. Archived at http://www.dieoff.com/page40.htm

11 . Garret Hardin, "An Ecoloate View of the Human Predicament," Global Resources: Perspectives and Alternatives. Archived at http://dieoff.org/page14.htm See also David R. Klein, "The Introduction, Increase, and Crash of Reindeer on St. Matthew Island," Archived at http://www.dieoff.org/page80.htm

12 . Corson, Walter H., The Global Ecology Handbook: What You Can Do About the Environmental Crisis, p. 25

13 . Kirkpatrick Sale, Dwellers in the Land, p. 24-26. Archived at http://dieoff.org/page14.htm

14 . Jared Diamond, "Easter's End," Discover Magazine (August 1, 1995). Archived at http://www.oilcrash.com/easter.htm and http://www.eces.org/articles/000543.php

15 . Ibid.

Notes
Pages 41 to 59

16 . Jared Diamond, "Why Societies Collapse," Delivered at
Princeton University on October 27th, 2002. Archived at
http://eces.org/articles/000544.php and
http://www.abc.net.au/rn/talks/bbing/stories/s707591.htm
27th, 2002. Archived at http://eces.org/articles/000544.php and
http://www.abc.net.au/rn/talks/bbing/stories/s707591.htm

17
. Michael Meacher, "Plan Now for a World without Oil"
Financial Times, (January 5th,
2004). Archived at
http://www.energycrisis.com/uk/planNow.htm

18 . Archived at
http://www.futuresedge.org/World_Population_Issues/top10_195
0_2050.html

19 . Ibid.

20 . David Pimentel and Mario Giampetro, "Food, Land,
Population and the US Economy"
Carrying Capacity Network, (November 21, 1994). Archived at
http://www.dieoff.com/page40.htm

21 . Ibid.

22 . "China's Rising Food Prices Could Signal Global Food
Crisis," Terra Daily □ (November 19,
2003). Archived at
http://www.terradaily.com/2003/031119092535.t70a5roc.html

120

Notes
Pages 41 to 59

23 . "The Greatest Story Never Told" American Assembler,
Archived at
http://www.americanassembler.com/issues/peak_oil/peak_oil.ht
ml

24 . Archived at http://www.hubbertpeak.com/de/lecture.html
and
 http://www.mbendi.com/indy/oilg/p0070.htm

25 . Lee Dye, "Dried Up? Are we Running Out of Oil?" ABC
News, (February 10, 2004).
Archived at
http://abcnews.go.com/sections/SciTech/DyeHard/oil_energy_dy
ehard_040211-
3.html

26 . Michael Ruppert, "Behind the Blackout: Bush Insider
Speaks," From the Wilderness,
(August 21, 2003). Archived at
http://www.fromthewilderness.com/free/ww3/082103_blackout_
summary.html

27 . William F. Engdahl, "Iraq and the Problem of Peak Oil,"
Current Concerns, (January 26,
2004). Archived at
http://www.currentconcerns.ch/archive/2004/01/20040118.php

Notes
Pages 41 to 59

28 .As quoted by Larry Everest, Oil, Power and Empire.
Archived at
http://www.counterpunch.org/everest12132003.html

29. Michael Ruppert, "Paris Peak Oil Conference Reveals
Deepening Crisis" From the
Wilderness, (May 30, 2003). Archived at
http://www.fromthewilderness.com/free/ww3/053103_aspo.html

30. Graham James. "World Oil and Gas Running Out," CNN
International, (October 2, 2003).
Archived at
http://edition.cnn.com/2003/WORLD/europe/10/02/global.warmi
ng/

31 . Joseph A Giannone, "Major Oil Stocks Fall as Shell Revises
Reserves," Reuters Business
News, (January 9, 2004). Archived at
http://uk.biz.yahoo.com/040109/80/eiqr3.html

32 . "El Paso Trims Reserves 41%" Reuters (February 13, 2004)
Archived at
http://www.reuters.com/financeNewsArticle.jhtml?storyID=4378
340&type=bondsNews

33 . Jeremy Rifkin, "The Perfect Storm That's About to Hit,"
The Guardian, (March 24, 2004).

122

Notes
Pages 41 to 59

34 . David Good, "Will Shortfalls in World Grain Production Continue?" Farmdoc, (November 17, 2003). Archived at http://www.farmdoc.uiuc.edu/marketing/weekly/html/111703.html

35 . Archived at http://www.terradaily.com/2003/031119092535.t70a5roc.html

36 . Richard Heinberg, "Oil and Gas Update," Museletter Number 142, (January 2004)

37 . See http://news.bbc.co.uk/2/hi/americas/3560433.stm

38 . See http://www.peopleandplanet.net/doc.php?id=668 and http://www.prb.org/Content/NavigationMenu/PRB/Educators/Human_Population/Populatio n_Growth/Population_Growth.htm)

39 . Archived at http://www.fromthewilderness.com/free/ww3/04_04_02_oil_recession.html

40 . Matthew R. Simmons, "The World's Giant Oilfields," M. King Hubbert Center for Petroleum Supply Studies, Colorado School of Mines, (January 2002). Archived at http://www.currentconcerns.ch/archive/2004/01/20040118.php

Notes
Pages 41 to 59

41. Lita Epstein, C.D. Jaco, and Julianne C. Iwersen-Neimann,
The Politics of Oil, p.12

42 . Archived at
http://www.earthisland.org/eijournal/new_articles.cfm?articleID
=713&journalID=69

43 . "Petro Canada Reviews Oil Sands Strategy" Rigzone, (May
2, 2003). Archived at
http://www.fromthewilderness.com/free/ww3/062303_nat_gas_c
risis.html

44 . "New Sources as Oil Supplies Grow, US Is Less Reliant on
the Middle East," ASPO-ODAC
Newsletter no. 15, p. 11 (March 2002).

45 . Energy Information Agency, Annual Energy Outlook 1998
with Projections to 2020, p.17.

46 . http://www.wolfatthedoor.org.uk/

47 . C.J Campbell and Jean Laherrere, "The End of Cheap Oil?"
Scientific American, (March
1998). Archived at http://www.dieoff.org/page40.htm

124

Notes
Pages 41 to 59

48 . William Engdahl, "Iraq and Problem of Peak Oil," Current Concerns, (March 16, 2004).
Archived at
http://www.currentconcerns.ch/archive/2004/01/20040118.php

49 . Michael Meacher, "Plan Now For a World without Oil," Financial Times, (January 5, 2005).
Archived at http://www.energycrisis.com/uk/planNow.htm

50 . Thom Hartmann, "The Empire Needs New Clothes," Common Dreams, Archived at
http://www.commondreams.org/views03/0311-07.htm

51 . Chris Skerbowski, "Oil Field Mega Projects, 2004," Petroleum Review, (January 2004)
Archived at http://www.odac-info.org/bulletin/documents/MEGAPROJECTSREPORT.pdf

52 . Ibid

53 . Ibid

54 . Carola Hoyos, "A Battle Against Shrinking Reserves," Financial Times, (January 7th, 2004)

55 . C.J Campbell and Jean Laherrere, "The End of Cheap Oil?" Scientific American, (March
1998). Archived at http://www.dieoff.org/page40.htm

Notes
Pages 41 to 59

56 . C.J. Campbell, Association for the Study of Peak Oil Newsletter, no. 15 (March 2002).
Archived at http://www.asponews.org/ASPO.newsletter.015.php

57 . http://www.wolfatthedoor.org.uk/

58 . "Interview with Matthew Simmons," From the Wilderness, (August 18, 2003). Archived at http://www.oilcrash.com/blackout.htm

59 . Dr. David Goodstein, Out of Gas The End of the Age of Oil, p. 35

60 . Albert A. Bartlett, "An Analysis of US and World Oil Production Patterns Using Hubbert-
Style Curves," Mathematical Geology, vol. 32, no. 1 (January 2000)

61 . Interview with Dr. Colin Campbell, Archived at http://globalpublicmedia.com/INTERVIEWS/COLIN.CAMPBELL/
 62 . Oliver Morton, "Fuel's Paradise," Wired Magazine, Archived at http://www.wired.com/wired/archive/8.07/gold.html

63 . Michael Ruppert, "Colin Campbell on Oil," From the Wilderness, (October 10, 2002)
Archived at http://www.fromthewilderness.com/free/ww3/102302_campbell.html

126

Notes
Pages 41 to 59

64 . Lita Epstein, C.D. Jaco, and Julianne C. Iwersen-Neimann, The Politics of Oil, p.22

65 . Dr. Colin Campbell, Association for the Study of Peak Oil Newsletter, no. 35 (November 2003). Archived at http://www.asponews.org/HTML/Newsletter35.html

66 . Jay Hanson, "Energetic Limits to Growth," Energy Magazine (Spring 1999). Archived at http://dieoff.org/page175.htm

67 . Matthew Simmons, "Revisiting Limits to Growth: Could the Club of Rome Have Been Correct?" Archived at http://www.greatchange.org/othervoices.html

68 . Richard Duncan, "Three World Oil Forecasts Predict Peak Oil Production," Oil and Gas Journal (May 26, 2003), as quoted by Richard Heinberg, "Oil and Gas Update," Museletter No. 142 (January 2004)

69 . See http://www.asponews.org/ASPO.newsletter.015.php

70 . Michael Ruppert, "Paris Peak Oil Conference Reveals Deepening Crisis," From the Wilderness (June 9, 2003). Archived at http://www.fromthewilderness.com/free/ww3/053103_aspo.html

Notes
Pages 41 to 59

71 . Mark Townsend and Paul Harris, "Now the Pentagon Tells Bush: Climate Change Will
Destroy Us," The Guardian, (February 22, 2004). Archived at http://observer.guardian.co.uk/international/story/0,6903,115351 3,00.html

72 . David Price, "Energy and Human Evolution", Population and Environment, A Journal of
InterdisciplinaryStudies, Volume 16, No. 4. p.301-319, (March 1995). Archived at
http://www.dieoff.com/page137.htm#2

NOTES
Pages 60-93

250. "Energy needs, choices, and possibilities: scenarios to 2050", Shell special publication. 2001. In this scenario, only in Europe would energy-efficiencymeasures be needed in order for renewables to hit the target.

252. Gobal annual electricity use at ten terawatt years = 87,600,000,000 kilowatt hours. The number of kilowatt hours per kilowatt of peak power of solar PV in a region as sumy as the Sahara would be 2,000. Therefore the kilowatts of peak power would be 43,800,000,000. 10 square metres delivers lkWp, which is 438 billion square metres, equal to a square 600 by 600 kilometres.

251. "Solar energy: brilliantly simple", BP pamphlet, available on UK petrol forecourts.

254. US Department of Energy f-igures quoted in Lester Brown, Eco-economy:
Building an Economy for the Earth. Earth Policy Institute, 2001.

255. Eco-economy, quoted at note 254.

256. For detailed argument see Arnon- Lovins, E. Kvle Datta, Odd-Even Busmes'
Jonathan G. Koomey and Nathan J. Glasgov', Winning tbe Oil Endgame:
lnnovation for Profit. Jobs, and Security, Earthscan. 2004, p. 240

257. /258 "Renewable energy: practicalities", Report of the House of Lords
Science and Technology- Committee, HL Paper 126-I, 2004

258. "You want the confidence to invest in renewable energy", Department of Trade and Industry information booklet, 2004.

NOTES
Pages 60-93

259.
"You want the confidence", quoted at note 258.'Thev say 20,000 homes, but use the average home consumption, which is of course energy inefficient.

260. "A microgeneration manifesto", Green Alliance, September 2004. The tens of thousands of small sites could make 500 megawafts to 2 gigawatts.

261. Anaerobic digestion is the breaking down ofcarbohydrates by bacteria in the absence of oxygen. Gasification entails heating wood chips on a fluidized bed": a controlled flow of air of steam. 'I'he result is a cornbustible mixture of gases, including carbon monoxide, hvdrogen and methane. Pyrolysis entails the heating of organic waste at high pressure, in the absence of air.
The result is a high-quality oil that can be used to fuel a power plant. These techniques are more eficient than traditional combustion because the gas, mixed with air,
burns at a higher temperarure, so driving the turbine more efficient.
262. Patricia Thornley, biomass expert on the UK Government's Renewables Advisory Board. Personal comrnunication.

263. Rather less electricity than in a conventional power plant.
264. See Power to the People, quoted at note 199 above.

265. John Gartner, "Automakers give biodiesel a boost", Wired, 23 September 2004.

130

NOTES
Pages 60-93

266. Information in this paragraph from "Field of dreams: is turning crops into fuels and chemicals the next big thing?". Economist, 7 April 2004.

267. Many fuel cells are of a "proton exchange membrane" (PEM) type. In these, a solid polymer is sandwiched berween two electrodes (an anode and a
cathode). A platinum catalyst stimulates hydrogen to give up an electron at the cathode.
.For further derails see
Hydrogen, Fuel Cells and the Prosperts for a Cleaner Planet. The MIT press, 2002.

268. In steam reforming, a hydrocarbon is first vaporized in a combustion chamber, and then flows into a steam reformer chamber, where a catalst breaks apart the gases and water vapour. Scientific American, May 2004, pp.66 71.

269. Electrolysis involves passing an electric current through water ro split the
molecules. Oxygen atoms are artracted to the anode (positive terminal) and hydrogen atoms are attracted to the cathode (negative terminal).
270. Questions about a hydrogen economy", quoted at 268.

271. Craig Simons, "The high road: if China steers its auto industry toward hybrids and perhaps hvdrogen cars the world may have no choice but to follow", Newsweek, 6-13 September 2004.
272.Winning the Oil Endgame, quoted at note 256.

NOTES
Pages 60-93

273. The Conde Nast Building in Manhattan is a prominent,Anerican exarnple. In the UK, Woking Borough Council's innovative low-carbon programme includes a large fuel cell at a sports complex.

274. Publicity pamphlets of Ovonic Batteries, a subsidiary of Energv Conversion Devices.
275. "Hybrid furure: Sky high prices are driving a quest fbr eftlcient energy"',
Newsweek, 8 September 2004.

276. Winning the Oil Endgame, quoted at note 256.
277. "The sustainability of higher oil prices",at note 84.
278. Nelson Schwartz, "Poor little rich company: on the back of $55 oil' Fortune,18 April 2005.

279. "Hubbert's Curve",quoted at note 60, pp. 56-61 (interview with M. King Hubbert).
280. "We can move towards a near-zero carbon furure, says Margaret Beckett"' www.strategy.gov.uk
281. Allan Jones MBE, "Woking: local sustainable energy community",presentation to Low Carbon Thames Gateway conference, Barking, l6 Junc2004.

282. Winning the Oil Endgome, quoted at note 256.

2Bl. Robert Monks, The New Global lnvestors, Capstone Press, 2001.
284. Clalton Christensen, Tbe lnnovator Dilemma: When New

Technologies Cause Great Firms to Fail. Harvard Business
School Press, 1997.

NOTES
Pages 60-93

286. Alex Kirby.', "When the last oil well runs dry", BBC News
Online, 16 April 2003

287. John McGaughey "Energy reserves: apocalypse now,
tomorrow or never?".
World Energy Review August 2001.

288. By hydrogen fuel plants I mean those using fossil-fuel
feedstock, not smallscale hydrogen-from-renewables plants.
289. Gordon MacKerrron
"Nuclear power and the characteristics of ordinariness
Energy Policy vol. 32, pp. 1957-1965.

290. Department of Trade and Industry - Creating a Louw
Carbon Economy,The Stationarv Office 2003 pp 6l-62.
296. A tale of two planets.
297. John Carey,
"Gloobal Warming why business is taking it so seriously"'
Business Week 30 August 2004

298."Fortune magazine backs renewable enerpy"' 16 August
2004
www.solaraccess com

299.New Energies Invest

300. Photon March 2004. The oil index quoted is the OXI.

NOTES
Pages 60-93

301.Michael Rogol Shintaro Doi and Anthony Wilkinson'
"Sun screen Investment opportunity in solar power"' Credit
Lyonnais Securities Asra report, July2004

302."C02 capture and storage"' website maintained bv IFIA
Greenhouse R&D Programme, http://wuw
co2sequestration.info/research_prograrnmes.htm

303. Mark Clayton America's new coal rush:
Christian Science Monitor 26 Februarv 2004.

304. Karen Armstrong. The Battle for God Harper Collins 2001

306. Information in this paragraph fiorn "Carbon sequestration –
technology roadmap and program plan, 2004", US Department of
Energy and National
Energy Technology Laboratory April 2004.

307. Vanessa Houlder, "The case for carbon capture and
storage", Financinl Times' 23 Januarv 2004
308. Nicola Jones, "Bubbling under", New Scientist Breaking
News, 20.June 2001.

309. Information in this paragraph from Maggie McKee, "Seas
absorb half of carbon dioxide pollution", New Scientist, 15 July
2004.
310. IPCC figures from the Third Scientist Assessment Report,
2001, quoted at note 143

311. James Randerson,"Forest experiment questions greenhouse
gas strategy", New, Scientist Breaking News, 15 April 2002.

www.ingramcontent.com/pod-product-compliance
Lightning Source LLC
Chambersburg PA
CBHW030758150426
42813CB00068B/3237/J